ROBOTICS IN EDUCATION:

an information guide

by

Veronica Sexauer Pantelidis

The Scarecrow Press, Inc.
Metuchen, N.J., & London
1991

British Library Cataloguing-in-Publication Data Available

Library of Congress Cataloging-in-Publication Data

Pantelidis, Veronica S.
Robotics in education : an information guide / by Veronica Sexauer Pantelidis.
p. cm.
Includes bibliographical references and index.
ISBN 0-8108-2466-3 (alk. paper)
1. Robotics--Study and teaching. I. Title.
TJ211.26P35 1991
629.8'92'071--dc20 91-24541
CIP

Manufactured in the United States of America

Printed on acid-free paper

To Nick and Nick, Jr.

who, as husband and son,

supported and participated in

this venture.

CONTENTS

PREFACE

This is a book on robots. For the purposes of this work, a robot is considered to be a robot if the author of the work cited calls it a robot, and if it is a mobile device that can perform work, even a simple task, and possibly do work usually considered the domain of human workers.

The goal of this book is to provide teachers with a source of information on robots, both in order to teach with them and to teach about them, so that teachers can prepare their students for the world that lies ahead, a world perhaps dominated as much by robots as our world today is dominated by the computer.

Robots today seem to be about where computers were in the 1950's. In those years, some companies were using computers. They were expensive. Quite a bit of basic research was being done. There were frequent new developments. And there were great expectations about what computers would accomplish in the future. Substitute the word *robot* and most of these descriptions would fit the robot of today, particularly robots used in industry and in service functions such as aiding the handicapped. Research in the field of computer science brought about the microcomputer which revolutionized many of our ways of doing things. Research in the field of robotics will one day soon result in another revolution, this time in the way humans are helped in manual rather than computing and calculating tasks. Such a revolution is inevitable. The task for the educational system is to prepare young people for the enormous changes in our lives that this revolution will bring about.

Beginning with an introduction, this book is divided into ten sections of annotated citations of materials on robots or with robots as a dominant theme, followed by six appendixes.

Section 1 covers reference sources on robotics such as dictionaries, encyclopedias, and online computer bibliographic databases.

Section 2 includes materials on the history of robots, general information about robots, and sources of career information.

Section 3 covers materials on the use of robots in schools and on robots designed specifically for educational use. It also includes materials on classroom activities and annotations of stories involving robots.

Section 4 deals with robots which can be used in the school setting but which are not considered as specifically for educational purposes, including personal, hobby, and toy robots. Information on building robots is also included in this section in the hobby robots, robot building grouping.

Section 5 covers uses of robots in fields other than education such as industry, agriculture, medicine, space, sports, and underwater.

Section 6 covers materials dealing with the social implications of robot use and how they may change our world.

Section 7 deals with the economic implications of robots.

Section 8 covers the impact of the robot on the worker in the workplace.

Section 9 considers materials dealing with the physical implications of the robot, especially safety factors.

Section 10 covers nonprint media, such as computer programs, filmstrips, and videorecordings, that can be used in teaching about robots.

Appendixes list robot associations that are of interest to the teacher, periodicals dealing with robotics, particularly the less technical ones, educationally useful robots, arranged by type, addresses of manufacturers and distributors of educationally useful robots, sources of electronic and other parts for building robots, and addresses of nonprint media sources for items covered in Section 10.

This sourcebook is selective, listing only a portion of the available periodical articles, books, and nonprint items on robots. The materials covered were chosen on the basis of their perceived value to the teacher, and many are grouped according to the level for which they seem most appropriate, using the broad designations preschool, elementary grades, and secondary grades.

In order to list fairly current materials in a rapidly changing field, most of the items included, except those of a strictly historical nature and fiction, have been published or produced since 1980. Highly technical treatments of robots and robotics are excluded.

An attempt has been made to list only materials that can be obtained fairly easily. Most materials listed should be available to the teacher either in the school media center, the local public library, or through interlibrary loan. For the most part, materials which would be either hard to locate or difficult to obtain via interlibrary loan have been excluded. All but a handful of the materials have been personally seen and read by the author.

Works were excluded that contain adult-level photographs or material which might be considered inappropriate for students even at the 12th grade level. For that reason, the more violent science fiction, as well as some authors whose fiction might be considered to be for a mature audience, has been omitted.

Citations within a subject category are arranged alphabetically by author. Works with no given author are arranged alphabetically by the first word in the title. Articles such as a, an, and the are not ignored when they are the first word in a title. All works with no author given are grouped before all works with an author given.

The index covers all citations and major topics in the annotations but does not cover the appendixes since these are each arranged alphabetically. Numbers next to entries in the index refer to citation numbers.

If the cover of a periodical shows a picture of a robot, this is noted in the citation annotation for the article in that periodical, in the belief that such information might be helpful to the teacher who is gathering pictures of robots for illustration in a lesson or for creating a display. Covers are indexed under the word "cover."

The robot of today is beginning to serve as a human assistant rather than a replacement. The robot of tomorrow may have civil rights and a social position in our society. It is the hope of the author that this book will help teachers prepare young people for that challenging tomorrow.

Veronica Sexauer Pantelidis, PhD
Department of Library and
Information Studies
East Carolina University
Greenville, NC 27858-4353

December 1990

ACKNOWLEDGMENTS

This book came into being as a result of several years of giving robotics programs and demonstrations at schools, conferences, meetings, and even a movie theater film debut, always assisted by my son, Nick, who started helping with the demonstrations when he was seven years old. He kept the programs going smoothly, fielded questions skillfully, and made the robots do his bidding with ease. To him I owe the greatest debt.

My husband, Nick Sr., left his business many times to drive us to program destinations. His support meant that I was able to give full attention to the program without transportation worries.

Ms. Emily Boyce, past chair of the Department of Library and Information Studies, East Carolina University, gave me every possible encouragement and acted as a catalyst for many of the ideas Nick Jr. and I had about programs and robot uses. Her close involvement with schools and support of our robot programs helped lead to the idea of a need for a book of this type.

Dr. Lawrence W. S. Auld, chair of the Department of Library and Information Studies, East Carolina University, has given me continual support, ideas, valuable critiques, and a broader view of robotics based on his keen insights of the world of the science fiction robot. He has also provided the introduction to this work.

Ms. Pat Guyette, interlibrary loan librarian at Joyner Library, East Carolina University, noted for her ability to track down elusive materials, has found such materials for me time after time, making my work much easier.

Also acknowledged is the support of Dr. Stan Garren of the Department of Manufacturing at East Carolina University, whose course on industrial robotics provided me with necessary background, and Dr. Barry DuVall and Dr. John Marshall of the same department, who gave me needed encouragement.

A number of graduate assistants have worked with me on this sourcebook over the years. I want to especially thank Linda Shoffner for her determined efforts to locate materials, Patty Chalovich for work on nonprint material sources, and Cynthia Solomon for work on the index, as well as Patsy Verreault and Imogene Turner.

I would also like to thank all the media specialists, teachers, conference program directors, and others who asked me to give robotics programs, made suggestions as to what they would like in an information sourcebook, and brought useful materials on robotics to my attention.

I would like to acknowledge the many students of mine who chose projects and wrote papers on the use of robots in education, and expressed interest in having an information sourcebook as a guide. Their interest in this topic has been a source of inspiration and motivation to me.

Dr. Diane Kester and Dr. Auld gave unlimited help and suggestions and quite a bit of their time as I produced the camera-ready copy of this manuscript using a desktop publishing program, and I owe them a debt of gratitude for their aid.

Finally I would like to thank my super friend and perfect next door neighbor, Marie Britt, who has listened and supported me through all the other things that have happened during the years I have been working on this guide, so that I kept working.

To all these people who helped to make this information guide a reality, I owe a great deal of appreciation.

V. S. P.

INTRODUCTION

Robot. The word brings to mind the idea of a machine in human form, and the robots of fiction have encouraged us to think in that way. But today, we have thousands of real robots that do not fit this image. Some are in our factories where they are employed in the manufacture of products such as automobiles. Others we take for granted in our daily lives though we seldom think of them as robots. But they are machines that do jobs that humans do, or, more often, used to do. For instance, consider the familiar red, green, and yellow robots that work at so many street corners. They have been common for most of a century. If there were a metal or plastic mannikin directing traffic through an intersection, we would recognize it immediately as a robot, and we would probably wonder from which movie it had escaped. Although the traffic light has shed its superfluous appendages to become more efficient, it remains a robot in its function. And when the traffic lights don't work, a human traffic officer is dispatched to act as a substitute. Or, look at the Automatic Teller Machine (ATM) outside your favorite bank. It is certainly convenient to be able to get money when the bank is closed, i.e., when the human employees have gone home and left the service operations of the bank in the circuits (we really can't say hands) of a robot.

We appreciate our robots and would give up few of them willingly or happily. We would like more of them in our lives as is indicated in this letter to the editor of *Micro Cornucopia*:

> ... but why have a robot project for such an inane thing as maze running? Don't misunderstand. I want a robot, and *now*. But a mere maze solver isn't worth the effort. I will build or buy a robot only if it will clean my bathroom.

> Let it be schedulable weekly for the porcelain, counters, and floors; monthly for the walls and ceiling. Let it do everything within the doorway except towels. I hope it isn't too expensive because I want it *now.*
>
> A.K. Holmes

Not content with labor-saving robots, both industrial and domestic, we imagine the possibility of robots as companions as in Tanith Lee's *Silver Metal Lover*, a beautiful love story with nuances that only more mature minds will fully appreciate. We also consider the theological implications of robots that can think and reproduce themselves in their own image. Norbert Wiener's *God and Golem* is only one of several discussions of the impact computers, and by implication robotics, might have on religious beliefs and practices. Whatever approach we choose to take, robots have captured our attention in many different ways.

Science fiction stories have had a great influence on our thinking about robots. Generally, they follow one of two formulas. The first depicts a society in which the robots have taken over, having relegated humans to some kind of irrelevancy, while the latter, not accepting this judgment, fight back with varying degrees of success or failure. The second depicts a society in which the robots do all of the work and the humans derive all of the benefits--we won't define either work or benefits, for there are some interesting reversals in some of these stories. The robots, existing in large numbers, are either pervasive in their presence or dispatch their responsibilities covertly. In either instance, the humans become inured to the presence of the robots, soon losing conscious awareness of them and taking them for granted just as we do with some of the robots around us.

One man who has given a great deal of thought to human-robot relations is Isaac Asimov, who promulgated his three laws of robotics in "Runaround," a short story first published in 1942. More recently, in *Robots and Empire*, he promulgated a fourth or Zeroth Law. These laws are:

> (0) A robot may not injure humanity or, through inaction, allow humanity to come to harm.

(1) A robot may not injure a human being, or through inaction, allow a human being to come to harm.

(2) A robot must obey the orders given it by human beings except where such orders would conflict with the first law.

(3) A robot must protect its own existence as long as such protection does not conflict with the first or second law.

From the human point of view, these four laws seem ideal, permitting maximum use of robots while protecting us from a robotic revolution, for we want robots that we can control, not robots that control us. But, would free-thinking robots consider them ideal? Perhaps they would propose an alternate set of laws beginning with "A human may not injure a robot, or through inaction, allow a robot to come to harm"?

Robots are fascinating, but they are not easily defined. Until this century, they existed largely in imagination. The term "robot" became popular after Karel Čapek's play, *R.U.R. (Rossum's Universal Robots)*, premiered at the National Theater in Prague, Czechoslovakia, in 1921. This dark tale of robotic revolt ends with the human masters being killed. The word robot, coined by Čapek's brother, Josef, who first used it in "Opilec," a short story published in 1917, comes from the Czech words *robotit*, which means "to drudge," and *robota*, which means "forced labor" or "servitude." Taking a lead from Čapek, Asimov coined the word "robotics" for the study of robots and their applications in "Strange Playfellow," a short story published a generation later in 1940.

The presence of robots on stage in 1921 was by no means the first appearance of these man-made creatures with various human attributes. The Greeks employed automata, human figures with mechanically controlled, often steam powered movements, in some of their temples. A mythical rabbi in Prague modelled a human figure in clay and breathed life into it, creating a golem which eventually ran amok. Dr. Frankenstein, with help from Mary Shelley, fashioned and brought to life the well-known monster which has since starred in so many movies. Mechanical dolls are primary characters in Delibes's ballet, *Coppélia*, and Offenbach's opera, *The Tales of*

Hoffmann. We are accustomed to all sorts of artificial beings in the movies, on stage, and on television.

What is a robot? We have seen already that they are not necessarily machines in human shape. Many of those in factories are patterned on only a portion of the human anatomy such as a mechanical arm and hand. As we have seen with traffic lights, the fact that, for the most part, contemporary robotic devices don't mimic human form doesn't make them any less robots. On the contrary, it emphasizes certain fundamental differences between robots and humans.

What, then, is a robot? The Robotics Institute of America would limit the term to any device that links intelligence to the capacity to do work. Dr. Pantelidis has chosen a broader definition--if a device moves and can do work, even of a limited type, it is possibly a robot. This latter characterization allows the inclusion of many ideas and concepts that would otherwise be excluded.

An all-encompassing definition of robot must take into account at least six different approaches to the problem of building such a device. Not all of the following represent robots per se, but, by giving thought to each of them, we can better understand what a robot is.

> The first approach is to assemble a machine which, to some degree, looks, moves, and behaves like a human. This machine may be based on a simple set of gears and levers, or it may be a much more complicated mechanism employing electronic circuitry and motors to drive the gears and levers. This is the classical approach and includes the previously mentioned balletic and operatic dolls.
>
> The second approach, a subset of the first, is to use only the body portion(s) necessary to perform a particular task or set of instructions. Most industrial robots, as mentioned above, belong to this category, consisting of, for example, an "arm" or "hand" that is designed to do one task or a particular type of task. These are an abstraction of our original idea of a robot.

The third approach is to grow a biological being which is composed of living cells, all of which were manufactured and developed in a laboratory. This artificial construction or cyborg is made of living matter, but it is not human, since it lacks human parents and birth. Lt. Data, the android serving on the U.S.S. *Starship Enterprise*, probably belongs to this class.

The fourth approach is to build another kind of biological being which is composed of both human and laboratory-grown parts. This artificial construction is an ambiguity, being endowed, in part, with human ancestors. Typically, the human portion would include the brain. Helva, the central character/starship in *The Ship Who Sang*, is a poignant example of this category.

The fifth approach does not produce a proper robot, but it is appropriate that it be included here. This is the "bionic" person with multiple body parts that have been replaced with much more powerful artificial parts. As more and more artificial parts are used as replacements, does this person eventually cease to be a human? If so, at what point does this person become a robot? Now, go back and ask the same kinds of questions about our manufactured being in number 4 above. We still see television reruns of "The Bionic Man," sometimes in the company of his romantic interest, the bionic woman.

The sixth approach is to create a robotic mind or virtual robot inside a computer. There are no external parts that mimic humans, but the reactions and thoughts are often represented as particularly human. Classic examples from this group are Harlie, the computer in *When Harlie was One*, who invents a huge make-work project so he won't be turned off, and Mike who was instrumental in coordinating the independence movement on the moon as described in *The Moon is a Harsh Mistress*.

Of course, the boundaries between these approaches are somewhat blurred when we encounter fictional creations that defy easy categorization. For example, the positronic brains with which the robots in many of Asimov's novels are equipped seem to fall into category 6, while their bodies might belong in category 1 or, possibly, category 3.

A further complication is sometimes referred to as a knowledge robot (e.g., a Knowbot™), a sort of robotic mind that is envisioned as a nimble expert system. With an expert system, we don't stop to worry whether our computer--or robot--can think. The special knowledge of a human expert is distilled into a computer program which has the appearance of mimicking the thought processes of the human expert in the particular area of specialization. Essentially, it is (1) an elaborate data base containing "everything" the human expert knows on the subject, and (2) cleverly designed programs with which the data base can be accessed and used. As long as the queries fall within the subject area of the data base and are within the capacity of the programs, a knowledge robot can produce striking results, but any query that deviates from the defined subject area and programmed capabilities will fail to be answered in a meaningful way. Although knowledge robots, strictly speaking, are not robots, our understanding of the latter will be enlarged by a consideration of the former.

As we find more and more robots in our lives, it is essential that we understand the purposes for which they are constructed as well as their capabilities and weaknesses. With such an understanding, we may succeed in living comfortably and securely in partnership with robotic helpmates. But, without such an understanding, the robots may go their own ways, losing interest in working for and coexisting with humans.

Yes, the world of robots is fascinating, and what better, safer place to explore these manufactured beings than in books and journals. Somewhere in the vast literature about them there is a robot for every taste and purpose. It is not feasible to cite them all in one work. In identifying those robots that are most useful in education, Dr. Pantelidis has chosen a selection of writings which includes robots of all kinds that are suitable for reading by children and young adults. In so doing, she has prepared a bibliography which can assist us in our effort to understand robots better and lead us and robots toward common interests and, even, a sort of friendship.

Lawrence W. S. Auld, PhD
Chairperson
Department of Library and
Information Studies
East Carolina University
Greenville, NC 27858-4353

References

Asimov, Isaac. *Robots and Empire.* Garden City, NY: Doubleday & Company, 1985.

Asimov, Isaac. "Runaround." *Astounding Science Fiction*. 29 (March 1942). Reprinted in *I, Robot*. New York: Gnome, 1950.

Asimov, Isaac. "Strange Playfellow." *Super Science Stories*. 1 (September 1940). Original title: "Robbie." Reprinted in *I, Robot*. New York: Gnome, 1950.

Čapek, Josef. *Lelio*. V Praze: Nákl. K. Neumannové, (1917) (Knihy dobrých autourú, 142)

Čapek, Karel. *R.U.R. (Rossum's Universal Robots); a Fantastic Melodrama in Three Acts and an Epilogue.* English version by Paul Selver and Nigel Playfair. New York: Samuel French, 1923.

Gerrould, David. *When Harlie was One; A Novel.* New York: Nelson Doubleday, 1972.

Harkins, William E. *Karel Čapek*. New York: Columbia University Press, 1962.

Heinlein, Robert A. *The Moon is a Harsh Mistress*. New York: Putnam, 1966.

Holmes, A. K. [letter to the editor] *Micro Cornucopia*. no. 52 (March-April 1990).

Lee, Tanith. *The Silver Metal Lover.* New York: Daw Books, 1981.

McCaffrey, Anne *The Ship Who Sang*. New York: Walker, 1969.

Wiener, Norbert. *God and Golem, Inc.; A Comment on Certain Points Where Cybernetics Impinges on Religion.* Cambridge: The MIT Press, 1964.

SECTION 1. REFERENCE SOURCES

Because robotics as an object of study is relatively new, a large collection of reference materials is not yet available. Moreover, the field of robotics is dynamic and rapidly changing so that reference materials quickly become out of date. Often the only materials available are so technical as to be incomprehensible to the layman. This lack of materials will change as robots assume a greater role in our society, and the user of this sourcebook should use these listings as a guide to the types of materials to look for and publishers who may be likely to publish related reference materials in the future. New or revised editions of materials listed should also be looked for and used whenever available.

This section includes materials which can be classified as reference sources. The books listed are generally of a less technical nature than those which would be used by robotics manufacturers and commercial robot users. The online computer bibliographic databases and electronic bulletin boards listed are those considered useful to the teacher and are not inclusive of all that are available.

BIBLIOGRAPHIES

1. Chin, Felix. *Automation and Robots: A Selected Bibliography of Articles*. Monticello, IL: Vance Bibliographies, 1981. 26 p.

   An annotated bibliography, covers articles from 1971 to 1980, primarily on commercial uses of automation, and industrial robots.

2. Chin, Felix. *Automation and Robots: A Selected Bibliography of Books*. Monticello, IL: Vance Bibliographies, 1982. 19 p.

   Covers primarily materials written during the 1970's on automation, automation of libraries, industrial robots, and industrial automation.

3. Garoogian, Andrew. *Robotics, 1960-1983: An Annotated Bibliography*. Brooklyn, NY: CompuBibs, 1984. 119 p.

   Covers general information, industry, the work force, educational applications, personal/home applications, other applications, role of government, research, robotics in Japan, Great Britain, Soviet Union, and other countries, education, training, and careers in the United States, and the future. Lists organizations, periodicals and newsletters, and personal robot producers.

4. Gomersall, Alan, and Farmer, Penny. *Robotics; An International Bibliography with Abstracts*. Bedford, England & Berlin: IFS (Publications), Ltd. & Springer-Verlag, 1984. 209 p.

   Covers general literature, research and industrial practice, design and performance, operating components and sensors, control theory and systems, and applications, for 1981-1983, with some citations for 1984. Well illustrated with photographs.

5. Kissinger, Pat. "Artificial Intelligence, Robots and Education: Selected Sources." *Education Libraries* 12 (Fall 1987), 81-83.

   Provides annotations on books, ERIC publications, and periodical articles dealing with artificial intelligence and robots and education.

6. McFadden, Thomas. *A Selected Bibliography on Robotics*. Vol. 1: *1980-March 1983*. Rochester, NY: Rochester Institute of Technology, Technical and Education Center of the Graphic Arts, 1983. 14 p.

   A bibliography on robotics compiled from *Graphic Arts Literature Abstracts* and other sources. Most entries are concerned with industrial applications.

7. Young, Kay. *Robotics, 1970-1983.* Monticello, IL: Vance Bibliographies, 1984. 93 p.
Covers journal articles from 1970 through 1983, books, audiovisual materials, directories, online data bases, market data information sources, and documents, and provides a list of periodicals focusing on robotics.

Juvenile Level

8. "Book Boosters; Robots." *Learning* 10 (December 1981), 51-54.
An illustrated bibliography of books on robots, space, and the future, written at an elementary level. Illustrated with drawings.

DICTIONARIES

9. Burger, Erich, and Korzak, Gunter. *Dictionary of Robot Technology in Four Languages: English, German, French, Russian*. Amsterdam: Elsevier Science Publishing Company, Inc., 1986. 273 p.
Lists English term with German, French, and Russian equivalents on same page, and provides separate German, French, and Russian indexes.

10. Rosenberg, Jerry M. *Dictionary of Artificial Intelligence and Robotics*. New York: John Wiley & Sons, 1986. 203 p.
Provides definitions for over 4000 terms. Commonly used symbols, acronyms, and abbreviations are also given.

11. Sheridan, Thomas B., and Smith, Bradford M. *Robotics and Artificial Intelligence: A Glossary.* Sacramento, CA: Business/Technology Books, 1984. 126 p.
An expanded version of Smith, et al., *A Glossary of Terms for Robotics*. Covers industrial robotics. Artificial intelligence

terms include computer vision and natural language processing. Arranged by areas, with an alphabetical index. Illustrated with drawings.

12. Smith, Bradford M.; Sheridan, Thomas B.; Albus, James S.; Barbera, Anthony J.; and VanderBrug, Gordon J. *A Glossary of Terms for Robotics*. Washington, DC: National Bureau of Standards, 1981. 87 p. Reproduced by National Technical Information Service, U. S. Department of Commerce, Springfield, VA.

    Covers terms used in the field of industrial robotics. Arranged by areas, with an alphabetical index. Illustrated with drawings.

13. Tver, David F., and Bolz, Roger W. *Robotics Sourcebook and Dictionary*. New York: Industrial Press, Inc., 1986. 258 p.

    Includes dictionary of types of robots, dictionary of applications, glossary, and computer-control terminology. Also provides information on manufacturers and typical specifications, with drawings, charts, and photographs of robots produced. Has drawings, charts, and photographs throughout. Primarily on industrial robots.

14. Waldman, Harry. *Dictionary of Robotics*. New York: Macmillan, 1985. 303 p.

    Concentrates on terms used in the field of industrial robotics. Includes drawings.

Juvenile Level

15. Radlauer, Ruth; Radlauer, Ed; Mather, Jean; and Mather, Bob. *Robot Tech Talk*. Chicago: Childrens Press, 1985. 63 p.

    A dictionary with extensive definitions, designed for young people. Covers all aspects of the field of robotics. Illustrated with drawings and color and black-and-white photographs.

DIRECTORIES

16. *Robotics, CAD/CAM Market Place 1985.* New York & London: R. R. Bowker Company, 1985. 239 p.

    Focuses primarily on industrial robotics. Includes sections on books, proceedings, conferences, and symposia, books and proceedings publishers directory, magazines/ serials, associations, educational institutes, research institutes, conferences and workshops, online databases, consultants, robot projects, robot manufacturers, and allied products.

17. Aleksander, Igor, consult. ed., and Schlesinger, Robert J., tech. consult. *The International Robotics Yearbook.* Cambridge, MA: Ballinger Publishing Company, 1983. 336 p.

    Contains a directory of manufacturers, suppliers and services, a directory of world research and development activities, sources of grants and funding available by country, and a list of robot associations by country.

18. Flora, Philip C., ed. *International Robotics Industry Directory.* 4th ed. Conroe, TX: Technical DataBase Corporation, 1984. 385 p.

    This edition of this annual includes information and photographs of industrial robots manufactured in the United States, England, France, Israel, Japan, Sweden, Wales, and West Germany, and information on robots manufactured in Switzerland. Provides information and photographs of robot parts and systems. Lists of consultants/systems houses, research institutes, and a glossary are included.

19. Flora, Philip C., ed. *Robotics Industry Directory, 1982.* La Canada, CA: Technical DataBase Corporation, 1982. 234 p.

    This edition of this annual includes information and photographs of industrial robots manufactured in the United States, and information on industrial robots manufactured in Japan. Also provides lists of British and other European robot associations, information and photographs of special application systems, computer-aided manufacturing (CAM) systems, and components and peripherals, lists of distributors, consultants, and research institutes, and a glossary.

20. Kogan Page, comp., with Aleksander, Igor, consult. ed. *The World Yearbook of Robotics Research and Development*. London & Detroit, MI: Kogan Page & Gale Research, 1985. 453 p.

    Covers research activities in universities, institutions, and companies around the world. Includes guide to grant awarding bodies, world index of research and development centers, world index of robotics researchers, and recent developments in robotics. Provides selected bibliography.

21. Kogan Page, comp., with Scott, Peter, consult. ed. *The World Yearbook of Robotics Research and Development*. 2nd ed. London: Kogan Page, 1986. 583 p.

    Covers aspects of robotics, current developments and national funding programs around the world, world directory of robotics research and development activities, world indexes of research and development centers and robotics researchers, subject index to research activities, and further information. Provides selected bibliography.

22. Rowe, William M. *Robotics Technical Directory*. Research Triangle Park, NC: Instrument Society of America, 1986. 170 p.

    Provides lists of consultants, information on industrial robot companies, with technical information, drawings, and photographs, a products and services index, listings of robotics research, robotics periodicals, and industrial robotics associations, and a glossary of terms.

23. Society of Manufacturing Engineers. *Directory of Robotics Education and Training Institutions in Colleges, Universities and Technical Institutes, 1984-85*. Dearborn, MI: Society of Manufacturing Engineers, 1984. 50 p.

    Lists institutions alphabetically and by state, and in Canada. Separately lists institutions offering robotic support courses, associate degree, bachelor degree, master's degree, and doctoral degree programs. Provides information on whom to contact and robotics laboratory availability for each institution listed. Includes information on whether there is a certificate awarded for robotic support courses.

ENCYCLOPEDIAS

24. Dorf, Richard C., and Nof, Shimon Y., eds. *International Encyclopedia of Robotics: Applications and Automation*. 3 vols. New York: John Wiley & Sons, 1988. 1304 p.

    Covers all types of robots, their components, characteristics and development, design and application, social impact and economic value, and new research and development. Illustrated with photographs and drawings.

HANDBOOKS

25. Gevarter, William B., ed. *Robotics & Artificial Intelligence: Four Overviews*. Orinda, CA: Business Technology Books, 1984. 696 p.

    Includes "Robotics: An Overview," "Expert Systems: An Overview," "Computer Vision: An Overview," and "Natural Language Processing: An Overview." Robotics section covers introduction, robot functions and features, robot vision, programming, kinematics and dynamics, research requirements, commercially available robots, components, vision systems, current robotics research, applications now and in the future, expanding robot population, and future directions. Illustrated with drawings.

26. Hunt, V. Daniel. *Industrial Robotics Handbook*. New York: Industrial Press, Inc., 1983. 432 p.

    Covers all areas of industrial robotics, including social and economic aspects. Provides information and photographs for current robotic systems. Includes sections on international robotic development, and the future of robotics, and a glossary, a bibliography, and information on institutions involved in robotic research and development. Illustrated with photographs and drawings.

27. Hunt, V. Daniel. *Robotics Sourcebook.* New York: Elsevier Science Publishing Company, Inc., 1988. 321 p.
Covers robot technology - the robotics industry, how robots work, and typical robot applications; industry competitiveness - United States industry performance, world class competitiveness, international competitors and their markets, and market trends and competitiveness; terminology; points of contact; industrial robot organizations; robot manufacturers, robot rental/lease firms, robot consulting/applications firms, and United States robotics research organizations; acronyms and abbreviations; bibliography; and robotic programs. Illustrated with drawings.

28. Luce, Lucia Corduan. *The Robotics Handbook.* Pittsburgh, PA: Westinghouse Electric Corporation, 1984. Variable paging.
Covers robots in industry and at Westinghouse, including Unimation robots.

29. Miller, Richard Kendall. *Industrial Robot Handbook.* Lilburn, GA: Fairmont Press, 1987. 686 p.
Covers industrial robots, the robot manufacturing cell, safety considerations, and applications in manufacturing, commerce, agriculture, health care, and other areas, and includes directories of robot and machine vision manufacturers.

30. Nof, Shimon Y. *Handbook of Industrial Robotics.* New York: John Wiley, 1985. 1358 p.
Covers all areas of industrial robotics, including social and economic aspects. Presents discussions on the impact of robotics on education and training. Illustrated with photographs and drawings.

ABSTRACTING AND INDEXING SERVICES

Printed Abstracts and Indexes

Microcomputer Index
Database Services, Inc.
Suite 1305
2685 Marine Way
Mountain View, CA 94043
Includes articles on robots and robotics which appear in the magazines indexed.

Robomatix Reporter
EIC/Intelligence
(Environment Information Center, Inc.)
245 West 17th Street
New York, NY 10011
Covers robots, their applications, and impact on society.

Robotics Technology Abstracts
Cranfield Institute of Technology
Cranfield, Bedford MK43 0AL, England
Covers robotics technology, including manufacture and use.

Online Computer Bibliographic Databases

Computer Database
Information Access Company
11 Davis Drive
Belmont, CA 94002
Available online through:
DATA-STAR, Data-Star Marketing, Limited, Plaza Suite, 114 Jermyn Street, London, SW1 6HJ, England;

DIALOG, DIALOG Information Services, Incorporated, 3460 Hillview Avenue, Palo Alto, CA 94304;
MDC, Mead Data Central (Information bank; NEXIS), 9333 Springboro Pike, Miamisburg, OH 45342.
Includes robotics.

Microcomputer Index
Database Services, Inc.
Suite 1305
2685 Marine Way
Mountain View, CA 94043
Available online through DIALOG, DIALOG Information Services, Incorporated, 3460 Hillview Avenue, Palo Alto, CA 94304.
Includes articles on robots and robotics which appear in the magazines indexed.

Microsearch
Information, Inc.
Suite 1414
1725 K Street, North West
Washington, DC 20006
Available online through ORBIT, System Development Corporation, SDC Information Services, 2525 Colorado Avenue, Santa Monica, CA 90406.
Includes articles on robots and robotics which appear in the magazines indexed.

RBOT Database
BRS Information Technologies
1200 Route 7
Latham, NY 12110
Available online through BRS.
A file on all aspects of robotics, 1980-1984.

Robomatix Online
EIC/Intelligence
(Environment Information Center, Incorporated)
245 West 17th Street
New York, NY 10011
Available online through DIALOG Information Services, Inc. and

ESRIN, Via Galileo Galilei (Rome), 00044 Frascati, Italy.
Covers robots, their applications, and impact on society.

Robotics Technology Abstracts
Cranfield Institute of Technology
Cranfield, Bedford MK43 0AL, England
Available online through DATASOLVE, Datasolve, Limited, 99 Staines Road West, Sunbury-on-Thames, Middlesex, TW16 7AH, England.
Covers robotics technology, including manufacture and use.

ELECTRONIC BULLETIN BOARDS

RoboBoard BBS
Phone 415-752-7228
San Francisco, CA
Maintained by The Robotics Society of America.

SECTION 2. GENERAL ORIENTATION TO ROBOTICS

Historically, there is evidence of robot-like devices or automata being produced at least 5,000 years ago. The early Egyptians and later the Greeks were apparently fascinated by such objects. This interest in the robot has continued through the centuries. Robots today are still considered by many as almost exotic objects, sometimes regarded with awe, often given personal names, and almost always arousing interest, especially in young people.

The tendency to create robots with features based on those possessed by humans is seen no more clearly than in the descriptions of robots given in most general information books. Robot arms, legs, touch, vision, and other features are almost always described, frequently with comparisons to corresponding human features. Much research also is being conducted with the express purpose of making robots more closely resemble human beings in ability, especially intelligence.

Section 2 includes materials discussing the history of robotics or considered as part of the history of robots and robotics, general information on robotics, information about the robot as an object (e.g. robot arms, hearing, voice), and career information for the student considering the field of robotics as a vocation.

Articles and books listed in the General Information - Robot Features grouping of this section will give the teacher much support material for class discussions on how robots are and are not like humans today and how they are likely to change in that respect in the future.

The number of items listed in the Career Information grouping shows that the field of robotics is considered a viable career. The field encompasses many areas, including social, economic, and physical impacts of robot use, as well as design, development, maintenance, and use of robots in industrial and nonindustrial applications. There is a wide variety of career options ranging from working with the handicapped to working underwater. Interested students should find the career materials helpful in describing this challenging field.

HISTORICAL BACKGROUND

31. "Demons and Monsters." *Time* 116 (December 8, 1980), 76-77.
    Briefly describes history of the robot and some of the stories and movies written about robots. Includes photographs.

32. Adcock, Craig. "Dada Cyborgs and the Imagery of Science Fiction." *Arts Magazine* 58 (October 1983), 66-71.
    Discusses the cyborg (part human, part machine) in fiction, film, and art. Illustrated with photographs and drawings.

33. Ayres, Robert, and Miller, Steve. "Robotic Roots Revisited." *Technology Review* 85 (May/June 1982), 39-42.
    Lists a chronology of significant developments in robotics technology from 3000 B. C. through 1982. Illustrated with photographs and drawings.

34. Boehn, Max von. *Puppets and Automata*. New York: Dover Publications, Inc., 1972. 257 p.
    The translated version of the second volume, *Puppenspiele*, of the original two-volume German work *Puppen und Puppenspiele*, first published in 1929. The first chapter on automata and movable images includes descriptions of early robots and automata.

35. Brumbaugh, Robert S. *Ancient Greek Gadgets and Machines*. New York: Thomas Y. Crowell Company, 1969. 152 p.

Describes many types of automated toys, theaters, and mechanical objects made in ancient Greece. Illustrated with photographs and drawings.

36. Čapek, Karel. "The Meaning of 'R. U. R.'" *The Saturday Review* 136 (July 21, 1923), 79.

Čapek explains that his interest in writing the play, *R. U. R.*, was in men not robots, and as a comedy of science and truth.

37. Chapuis, Alfred, and Droz, Edmond. *Automata; A Historical and Technological Study*. Translated by Alec Reid. Neuchatel, Switzerland & London: Editions du Griffon & B. T. Batsford, Ltd., 1958. 425 p.

Covers history of automata from early Egyptian times through the 1800's, with a section on automata and robots, particularly in the 1930's. Illustrated with color and black-and-white photographs and drawings.

38. Cohen, John. *Human Robots in Myth and Science*. New York: A. S. Barnes & Company, 1967. 156 p.

Discusses robots in antiquity, the Middle Ages, the 17th and 18th centuries, in the recent past, and in the immediate future. Also deals with robots in fiction, the motives of robot makers, and how robots are made. Illustrated with photographs, drawings, and portraits.

39. Frude, Neil. *The Robot Heritage*. London: Century Publishing, 1984. 252 p.

Discusses science fiction, films, and literary works featuring robots, using the thesis that the body of "humanoid" literature, the "robot heritage," can be of help in constructing a vision of the real robot future. Illustrated with photographs and drawings.

40. Hillier, Mary. *Automata & Mechanical Toys; An Illustrated History*. London: Jupiter Books, Ltd., 1976. 200 p.

Traces history of automata and mechanical toys from early Egyptian uses of 2300 B. C. to the 1970's, with special emphasis from the 18th century onwards. Illustrated with color and black-and-white photographs and drawings.

41. Jailer, Mildred. "The Fascinating Delights of Automata." *Hobbies* 91 (October 1986), 32-36.

    Surveys history of automata, particularly in the 18th and 19th centuries, and describes some of the automata produced during that period. Illustrated with photographs.

42. Keillor, G. "The Bangor Man." *New Yorker* 48 (October 14, 1972), 39.

    Describes early automatons, including Benjamin Franklin's "Philadelphia Man," and discusses the practice of using children to operate the inside mechanisms of mechanical animals used on tour.

43. Lund, Robert. "Made in Ohio by Robots." *Popular Mechanics* 134 (September 1970), 81-83, 204, 206, 208.

    Describes the use of Unimate robots at General Motors assembly plant, Lordstown, Ohio, a pioneer project in the automobile industry. Illustrated with drawings and photograph.

44. Mellow, James R. "Antiques: Automata." *Architectural Digest* 45 (December 1988), 140-145, 204.

    Discusses history of automata. Illustrated with color photographs of automata of the 19th and early 20th centuries.

45. Miller, Mark. "Tools and Monsters; the Robot in the Western Mind." *New Republic* 184 (May 16, 1981), 26-32.

    Discusses feelings, particularly fear, that people have had about robots throughout the ages.

46. Plank, Robert. "The Golem and the Robot." *Literature and Psychology* 15 (Winter 1965), 12-28.

    Traces history of the golem, Frankenstein's monster, and other robot-like creatures of literature, as well as the robot chess Turk and various automata, and their influence on the robot of science fiction.

47. Sheckley, Robert. "Masters and Slaves, Friends and Enemies - Tracking the Literary Robot." *New York Times Book Review* 90 (April 21, 1985), 40.

Discusses the robot in literature.

48. Walter, W. Grey. "An Imitation of Life." *Scientific American* 182 (May 1950), 42-45.

Discusses the results of experiments with two light-sensitive turtle robots. Illustrated with drawings.

49. Warrick, Patricia S. *The Cybernetic Imagination in Science Fiction*. Cambridge, MA: The MIT Press, 1980. 282 p.

Provides results of a study with the following objectives: to write a history of robots and computers in science fiction; analyze the history to discover recurring images, patterns, and meaning; to explore cross-fertilizations taking place between writers of science fiction in this area, and between the fiction and developments in theoretical science and technological innovation; and to make a critical judgment of the literary merit of science fiction by a study of cybernetic science fiction.

Fiction

50. Asimov, Isaac. *I, Robot*. New York: Gnome Press, Inc., 1950. 253 p.

A collection of nine short stories about robots, written by Asimov and published from 1941 through 1950, in *Super Science Stories* and *Astounding Science Fiction*. Includes "Robbie," published in 1940, under the title "Strange Playfellow," in which appears perhaps the first use of the word, "robotics," coined by Asimov. Also included is "Runaround," first published in 1942, in the science fiction magazine *Astounding Science Fiction*, in which the Three Laws of Robotics developed by Asimov are spelled out. The three laws are: (1) A robot may not injure a human being, or through inaction, allow a human being to come to harm; (2) A robot must obey the orders given it by human beings except where such orders would conflict with the first law; (3) A robot must protect its own existence as long as such protection does not conflict with the first or second law.

51. Čapek, Karel. *R. U. R.* in *R. U. R. and the Insect Play*, by The Brothers Čapek. Translated from the Czech by Paul Selver. London: Oxford University Press, 1961. 177 p.

    This edition containing the play *R. U. R.* is one of several books which includes the play in its contents. *R. U. R.*, standing for "Rossum's Universal Robots," was written by Čapek in 1921, and translated into English in 1923. It uses the word "robot," which is derived from a Czech word, "robota," meaning "forced labor" or "servitude," to describe a humanlike machine, manufactured and sold as cheap labor. Čapek's brother Josef first used the word in 1917, in his short story "Opilec." In the play, robots turn upon their human masters and destroy all but one. This human manages after many years to find two robots with nearhuman qualities who will begin a new "human" race.

52. Homer. *The Iliad*. Translated by W. H. D. Rouse. New York: New American Library, 1938. 309 p.

    In "Book XVIII," Thetis visits Hephaistos to obtain armor forged by the god for Thetis' son, Achilles. Hephaistos, who is lame, is helped by young women made of gold who can think, speak, use their muscles, and work, in other words, robots.

Films

53. Schaefer, Stephen. "Robots, the Men." *Film Comment* 22 (July/August 1986), 6-7.

    Surveys a number of films of the 1980's that feature robots.

Elementary Grades Level

54. Jennings, Gary. *March of the Robots; From the Manikins of Antiquity to the Space Robots of Today.* New York: The Dial Press, 1962. 152 p.

Although dated, about half the book deals with the history of robots, and descriptions of, and stories about, robots of the past. Illustrated with drawings.

GENERAL INFORMATION

55. "France Leaps Forward with Robots." *New Scientist* 87 (July 31, 1980), 369-370.

Describes first generation robots, programmed to carry out the same operation over and over, irrespective of work environment, and second generation robots, which have sensors that let them understand what is happening nearby, and their development and use at Renault, a French firm. Includes photographs.

56. "From Garbage Collection to Brain Surgery with Robots." *U. S. News & World Report* 95 (December 5, 1983), 63-64.

Reports on interview with Joseph F. Engelberger, president of Unimation, Incorporated. Discusses applications of robots, particularly in the future, the robot market, and impact of robots on employment. Illustrated with photograph of Engelberger and photograph of a bomb disposal robot.

57. "In the Land of Robots." *Business Month* 132 (November 1988), 67-69, 72, 74-75.

Excerpts from *Inside the Robot Kingdom* by Frederik L. Schodt. Describes Japanese industrial robot manufacturers, especially Fanuc and GMF Robotics. Illustrated with color photographs.

58. "Robotics: An Interview with Dr. Arthur Harkins." *Library Hi Tech* 1 (Summer 1983), 7-15.
   Thomas Surprenant interviews Arthur Harkins on robots, how they are used, and how they may be used in the future.

59. "Robots with Human Intelligence." *The Futurist* 23 (March/April 1989), 52-53.
   Presents ideas on a general purpose robot and predictions on robots in the future, taken from *Mind Children: The Future of Robot and Human Intelligence*, by Hans Moravec (see number 103). Illustrated with drawing.

60. "The Robot Exhibit." *American Craft* 44 (April/May 1984) 12-13.
   Reports on "first comprehensive exhibition on robots to be organized in the United States," held at the American Craft Museum at International Paper Plaza, New York City, January 13 - May 12, 1984, and curated by Robert Malone. Article provides definitions and robot chronology based on exhibition catalog. Includes photographs and drawings.

61. Aleksander, Igor, and Burnett, Piers. *Reinventing Man; The Robot Becomes Reality*. New York: Holt, Rinehart and Winston, 1983. 301 p.
   Discusses prospects of creating a robot with artificial intelligence comparable to human intelligence. Describes WISARD, unprogrammed computer capable of learning from experience and by induction.

62. Asimov, Isaac. "Books." *Omni* 8 (October 1985), 38.
   Discusses Asimov's Three Laws of Robotics, and the creation of a fourth, the Zeroth Law, which states "A robot may not injure humanity, or through inaction, allow humanity to come to harm." Illustrated with photograph.

63. Asimov, Isaac. "The Robot in the 21st Century." *Radio-Electronics* 58 (May 1987), 99-101.

Discusses the social impact of robots, re-educating society, and humans versus robots. Illustrated with color drawing and photographs.

64. Asimov, Isaac, and Frenkel, Karen A. *Robots; Machines in Man's Image*. New York: Harmony Books, 1985. 246 p.

Deals with history of robots, their creators, how they work, making them "smart," applications in industry, home, and promotion, hobby robots, effect of robots on the economy, and robots and society. Provides lists of manufacturers, publications, and associations. Includes bibliography, black-and-white and color photographs, charts, and drawings.

65. Barrett, F. D. "The Robot Revolution." *The Futurist* 19 (October 1985), 37-40.

Describes ways robots are being used, particularly in business and industry. Illustrated with photographs.

66. Begole, Christine. "What Robots Can Really Do." *Glamour* 83 (March 1985), 260.

Features Odex 1, the six-legged walker; Sammy, the robot piano player; RB5X, a robot able to help the disabled and the elderly; and the problems of making a robot. Illustrated with photographs.

67. Bell, John. "'We'll Have Robots Like Ladies Have Hats.'" *New Scientist* 97 (February 24, 1983), 528-529.

Summarizes an interview with Joseph F. Engelberger, a founder of Unimation, Incorporated, the first robot manufacturer. Gives Engelberger's views on how robots could change work and life in the United States and England in the future. Includes photographs.

68. Berger, Phil. *The State-of-the-Art Robot Catalog*. New York: Dodd, Mead, & Company, 1984. 148 p.

Covers mythological, fun (toy, show, personal/home), industrial, and future robots. Provides catalog and consumer information. Includes photographs.

69. Black, Kent. "The Well-Tempered Robot." *House & Garden* 161 (December 1989), 146-49, 172.
Discusses robots of the past, present, and future. Illustrated with color photographs.

70. Chen, Wayne. *The Year of the Robot.* Beaverton, OR: dilithium Press, 1981. 182 p.
Part I, "The Intellectual Robot," discusses the human traits of the robot and their application to the interpretation of human behavior, social phenomena, and the democratic institutions of the United States. Part II, "The Robotosyncrasies," a novel published in 1976 by Chen under the name Wayne Hawaii, tells the story of an engineer/scientist who has created a robot with human intelligence.

71. Cichowicz, Ron. "Robots: Who Wants Them and Why?" *Carnegie-Mellon Magazine* 2 (Winter 1984), 12-18.
Reports on interviews on robotics with researchers at Carnegie Mellon University, Pittsburgh, Pennsylvania. Illustrated with photographs.

72. Cohen, Charles L. "Robots Get Smart in Japan." *Electronics Week* 58 (January 21, 1985), 15-16.
Describes crawling, climbing, organ playing, and other robots being developed in Japan. Includes photographs.

73. Colligan, Douglas. "Robotic Soul." *Omni* 8 (June 1985), 67-70, 118, 121-122.
Reports on robots and robotic projects and research in Japan.

74. Compton, Michael M. *Understanding Robots*. Sherman Oaks, CA: Alfred Publishing Company, Inc., 1983. 47 p.
Discusses why robots are used, history of robots, what robots do, types of robots, robotic subsystems, robot control systems, robots that one can buy, and how they will be used in the future. Provides a section on where to find more information and a glossary. Illustrated with photographs and drawings.

75. Corcoran, Elizabeth. "Homebody." *Scientific American* 261 (October 1989), 77-78.

    Tells the story of Joseph F. Engelberger, who, with George C. Devol, built the Unimate industrial robot in 1961. Illustrated with photograph.

76. Cornish, Blake M. "The Smart Machines of Tomorrow." *The Futurist* 15 (August 1981), 5-13.

    Includes discussion of industrial robots, robot vision, and SPEDA, a personal robot built by Louis Steinberg of Huntington, New York. Illustrated with photographs.

77. Cusumano, Michael. "Robots Step Out of the Factory." *New Scientist* 97 (January 6, 1983), 31-32.

    Discusses use for robots in Japan, lists major robot manufacturers and the types of robots made, reports on robots being developed, and considers robotics in the future. Includes photographs and drawing.

78. Davies, Owen. *The Omni Book of Computers & Robots*. New York: Kensington Publishing Corporation, 1983. 413 p.

    Part Five, "Computers That Walk," includes the following reprinted *Omni* articles on robots and their applications: "Robots That Dust," by Richard Wolkomir, "Self-Reproducing Robots/Born Again Robots," by Robert A. Freitas, Jr., and "Japanese Robots," by R. Bruce McColm.

79. Deken, Joseph. *Silico Sapiens; The Fundamentals and Future of Robots*. New York: Bantam Books, 1985. 259 p.

    Discusses the evolution from computers to robots, covering three types of robots: "microbots" (the "intellification" of personal machines), "midrobots" (robot workers and human employment), and "macrobots" (reshaping the world environment with land robots called "geobots," with underwater robots called "hydrobots," and with space robots called "astrobots," in the new "robosphere"). Also considers micro-mindworks (the brain of the robot), how the robot moves, senses, and perceives, and whether robot intelligence endangers or enhances the human species. Provides a glossary. Illustrated with drawings.

80. D'Ignazio, Fred. "Robots That Roll, Crawl and Bounce." *Compute!* 6 (January 1984), 34-42.
Describes the Robotics Institute at Carnegie Mellon University, Pittsburgh, Pennsylvania. Discusses current robot projects and explains in detail the robot Rover, in size and shape a distant cousin of the *Star Wars* robot R2D2.

81. Edelhart, Michael. "Robots: Fantasies and Realities." *Omni* 5 (April 1983), 80-89.
Describes uses of robots as fantasy in entertainment, and as reality in industry and other areas. Illustrated with photographs. Cover of this issue shows picture of robot hand reaching out to touch human hand.

82. Eiji, Nakano. "Designing Tomorrow's Robots." *Japan Echo* 10 (Special Issue, 1983), 59-63.
Considers today's robotics technology, the second-generation robot and beyond, concepts of future working robots, and the human factor. Includes photographs, diagrams, and charts.

83. Engel, C. William. *The World According to Robo the Robot.* Hasbrouck Heights, NJ: Hayden Book Company, 1985. 242 p.
Considers history of robots, how a robot works, how it is programmed, applications, and the future. Written from the viewpoint of a robot. Includes diagrams, glossary, and suggested readings.

84. Engelberger, Joseph F. *Robotics in Service.* Cambridge, MA: The MIT Press, 1989. 248 p.
Covers introduction to robotics, robot anatomy, control systems, sensory perception, artificial intelligence, and service applications. Illustrated with drawings and color and black-and-white photographs.

85. Fjermedal, Grant. "Surrogate Brains." *Omni* 9 (October 1986), 38, 160.

Taken from *The Tomorrow Makers*, by Grant Fjermedal (see number 86). Reports on conversations with robot researchers around the world concerning the creation of computer copies of the human brain to achieve longer lives or immortality.

86. Fjermedal, Grant. *The Tomorrow Makers*. New York: Macmillan Publishing Company, 1986. 272 p.

Reports on research being done on robotics at Carnegie Mellon University, Massachusetts Institute of Technology, Stanford University, University of Tokyo, Waseda University (Tokyo), and other institutions, government organizations, and companies in the United States and Japan.

87. Fredkin, Edward. "Robotics." *Research & Development* 26 (June 1984), 242-249.

Discusses the question "How good a robot can we build with the technology we have?", robots that may be built in the future, and possible social and economic consequences. Illustrated with photographs and drawings.

88. Geduld, Harry M., and Gottesman, Ronald, eds. *Robots, Robots, Robots*. Boston: New York Graphic Society, 1978. 246 p.

An anthology of writings about robots. Categories include robots in legend and literature, machine art, automata, man and machine or man versus machine?, movies and machine men, and future vision. Provides a listing of all films which included androids, automata, or puppets, from 1897 through 1977. Includes photographs, drawings, and a bibliography.

89. Gleick, James. "Why Can't a Robot Be More Like a Man?" *The New York Times Magazine* (December 11, 1988), 52-54, 60, 116-118.

Discusses work being done to produce a general purpose robot for everyday use. Illustrated with color and black-and-white photographs.

90. Hall, Ernest L., and Hall, Bettie C. *Robotics; A User-Friendly Introduction*. New York: Holt, Rinehart and Winston, 1985. 254 p.

    Designed as an introductory text for undergraduate robotics courses. Covers an introduction to robotics, history of robots, robot components and operation, intelligent robot programming, robot sensors, applications of industrial robots, economic considerations and justification of industrial robots, social impact of industrial robots, responsible technology, and robot applications today and in the future. Provides a glossary. Illustrated with photographs and drawings.

91. Heisler, Gregory, and Witty, Susan. "The Robots Are Coming." *Life* 7 (December 1984), 169-172, 174, 176.

    Reports on a variety of uses for robots outside of industry. Illustrated with color photographs.

92. Helmers, Carl. *Robotics Age: In the Beginning*. Rochelle Park, NJ: Hayden Book Company, Inc., 1983. 241 p.

    Reprints a collection of articles on robotics from the periodical *Robotics Age*. Articles are arranged under the following headings: Power and Control; Interactions: Senses, Vision, and Voice; and Applications and Development. Illustrated with photographs and drawings.

93. Holland, John M. *Basic Robotic Concepts*. Illustrated by T. R. Emrick. Indianapolis, IN: Howard W. Sams & Co., Inc., 1983. 270 p.

    Considers motors, manipulators, mobility, and vision systems of robots. Discusses applications. Mathematical equations and sample computer programs are given in appendixes.

94. Hunt, V. Daniel. *Smart Robots; A Handbook of Intelligent Robotic Systems*. New York: Chapman & Hall, 1985. 377 p.

    Covers introduction to smart robots, artificial intelligence for smart robots, sensor-controlled robots, machine vision systems for robots, robot manipulators, natural language processing, expert systems for robotics, and computer integrated manufacturing. Includes bibliography and glossary. Illustrated with photographs and diagrams. Includes large number of tables.

95. Kelly, Derek. *A Layman's Introduction to Robotics*. Princeton, NJ: Petrocelli Books, 1986. 209 p.
Covers what robotics is, history of robots, robotic concepts, robotic hardware, software of robotics, implications of robotics, careers in robotics, and future of robotics. Has glossary and bibliography. Illustrated with photographs.

96. Kemp, Mark. "The Robot Reality." *Discover* 9 (November 1988), 67-70.
Discusses what robots can do, some research projects to develop robot capabilities, and what robots may be able to do in the year 2001. Illustrated with color photographs and drawings.

97. Krasnoff, Barbara. *Robots; Reel to Real*. New York: Arco Publishing, Inc., 1982. 154 p.
Discusses science fiction and real robots, industrial robots, robots in space, "show" robots used for advertising and fun, "homemade" robots, and the future. Includes resource list. Illustrated with photographs and drawings.

98. Lamb, John. "Robots Plot Their Escape from Factory Boredom." *New Scientist* 118 (May 5, 1988), 40.
Reports on robot research being conducted in Great Britain and the United States.

99. Lecht, Charles. "Robotics a la Bradbury: Imitations of Nature." *Computerworld* 17 (June 13, 1983), 67-68.
Contemplates forms that future robots may take, after considering the robot mouse, robot fleas, and other robots in Ray Bradbury's *Martian Chronicles*.

100. Logsdon, Tom. *The Robot Revolution*. New York: Simon & Schuster, Inc., 1984. 207 p.

Covers what a robot is, robots in history, machines that see, feel, and think, the factory of the future, robots and military power, show robots, how to build a robot, and robots for the future. Illustrated with photographs and drawings.

101. McDermott, Jeanne. "Robots Are Playing New Roles as They Take a Hand in Our Affairs." *Smithsonian* 14 (November 1983), 60-69.

Reports on research being done and new developments in the field of robotics. Describes and illustrates various applications of robots. Includes photographs.

102. Minsky, Marvin, ed. *Robotics.* Garden City, NY: Anchor Press/Doubleday, 1985. 317 p.

Contains articles by ten contributors on intelligent machines, addressing what robots do today, what they will do in the future, and the implications for humans.

103. Moravec, Hans. *Mind Children: The Future of Robot and Human Intelligence.* Cambridge, MA: Harvard University Press, 1988. 214 p.

Discusses the present and future evolution of human equivalence in machines such as computers and robots, in their capacity to reason, perceive, interact with, and change their environment. Illustrated with photographs and drawings.

104. Nash, J. Madeleine. "Ouch! Ouch! Cut It Out!" *Time* 129 (February 9, 1987), 78.

Describes robot displays at "Robots and Beyond," an exhibition of intelligent machines at the Boston Museum of Science. Includes color photographs.

105. Perrin, Noel. "We Aren't Ready for the Robots." *Wall Street Journal* 207 (February 25, 1986), 30.

Reports on uses of robots around the world.

106. Peterson, Ivars. "Run, Robot, Run." *Science News* 127 (February 16, 1985), 109.

Reports on personal and other robots shown at the Consumer Electronics Show held in Las Vegas, Nevada, January 1985. Illustrates a computer screen from the computer program *ChipWits*.

107. Prather, Richard D. "Universal Robotics Systems: Homebrew Developments in the Field." *Library Hi Tech* 1 (Summer 1983), 17-19.

Describes robot projects of Universal Robotics Systems in Palo Alto, California. Illustrated with photograph.

108. Regis, Ed. "Interview: Hans Moravec." *Omni* 11 (August 1989), 74-76, 78-79, 86, 88, 91-92.

Presents an interview with Hans Moravec concerning future robots, their form, and their capabilities. Illustrated with photograph.

109. Ringel, Susan S. "The Robot Lady." *Carnegie-Mellon Magazine* 7 (Spring 1989), 12-16.

Reviews the life and work of K. G. Engelhardt, director of the Health and Human Services Laboratory at the Robotics Institute, Carnegie Mellon University, Pittsburgh, Pennsylvania. Work is especially oriented toward robots for health care, vocational rehabilitation, the handicapped, and older people. Illustrated with photographs. Cover of this issue shows color photograph of K. G. Engelhardt and part of her toy robot collection.

110. Robillard, Mark J. *Advanced Robot Systems*. Indianapolis, IN: Howard W. Sams & Co., Inc., 1984. 215 p.

Covers moving-platform mechanics, software for rover systems, manipulator systems, hardware, software for manipulators, personal robot hardware and software, applications, and advanced techniques. Describes Hero 1 circuitry, senses, and programming in depth. Includes photographs, drawings, schematics, and software programs.

111. Robillard, Mark J. *Microprocessor Based Robotics*. Illustrated by R. E. Lund. Indianapolis, IN: Howard W. Sams & Co., Inc., 1983. 220 p.

Covers hand and arm mechanics, electromechanics, electronic sensors, computer control, and robot communication control. Describes Big Trak in depth. Provides lists of parts and robot suppliers, and a bibliography. Includes photographs, drawings, schematics, and software programs.

112. Rochester, Jack B., and Gantz, John. *The Naked Computer*. New York: William Morrow and Company, Inc., 1983. 335 p.

Chapter 8, "The Steel-Collar Work Force: March of the Robots," covers facts about robots, robots in films, and jobs for robots.

113. Rogers, Michael. "Robots Find Their Place." *Newsweek* 111 (March 28, 1988), 58-59.

Describes how robots are being used and how they may be used in the future. Illustrated with color photographs.

114. Safford, Edward L., Jr. *The Complete Handbook of Robotics*. Blue Ridge Summit, PA: TAB Books, 1978. 358 p.

Provides an introduction to robots, and covers robot systems, servomechanism systems, power sources, mobility, sensors, the robot brain, tips on hobby robot construction, radio control interfacing to a computer, commercial robots, factory robots, robot societies, and the future of robots. Illustrated with photographs, drawings, and schematics.

115. Safford, Edward L., Jr. *Handbook of Advanced Robotics*. Blue Ridge Summit, PA: TAB Books, Inc., 1982. 468 p.

Covers definitions, examples, and ideas, the domesticated android, use, operation, and construction of a radio-controlled robot, robot control programming, programming the Prab Versatran Robot, robot hands and arm kinematics, a computer-operated hobby robot, the ASEA robot system, robotic vision and speech, ranging systems, numerically controlled robotic machines, and miscellaneous ideas and discussions. Illustrated with photographs, drawings, and schematics.

116. Schodt, Frederik L. *Inside the Robot Kingdom*. Tokyo & New York: Kodansha International, Ltd., 1988. 256 p.
Reports on robots in Japan, covering what a robot is, the first Japanese robot, cartoon robots, toy robots, industrial robots, other types of robots, religion and robots, and future and experimental robots. Illustrated with photographs and drawings.

117. Schwaller, Anthony E. "What Is a Robot, Anyway?" *School Shop* 45 (September 1985), 27-29.
Provides information on the advantages of robots, robot terminology, and classification of robots. Includes illustrations.

118. Scott, Peter B. *The Robotics Revolution; The Complete Guide for Managers and Engineers*. Oxford, England: Basil Blackwell, 1984. 345 p.
Covers robotics background information, robotics technology, robot applications in industry, societal, organizational, and economic considerations, present research in applications both industrial and nonindustrial, and future possibilities. Includes glossary. Illustrated with drawings and photographs.

119. Shaver, Robert W. "Robots: Education's New Challenge." *School Shop* 42 (November 1982), 18-19, 21.
Reviews history of the computer and robot, the growth of industrial applications, the future, social implications, and education's role.

120. Solomon, Stephen. "Miracle Workers." *Science Digest* 89 (December 1981), 38, 40, 42, 44.
Provides a history of robotics and discusses what industrial robots can do and their applications. Includes photographs.

121. Stein, Kathleen. "Robo-Shock." *Omni* 5 (April 1983), 44-46, 48, 50-51, 90-91.

Discusses what robots can and cannot do, robotic research, problems, and future uses. Illustrated with color drawing.

122. Stokes, Henry Scott. "Japan's Love Affair with the Robot." *The New York Times Magazine* 131 (January 10, 1982), 24-27, 60-62, 64, 75, 80.
Reports on Japanese robots and the ways they are used in Japan.

123. Sullivan, Barbara A. "Robots: How They Will Affect Your Future." *Forecast on Home Economics* 29 (October 1983), 33-37.
Discusses robots in the home, in school, in medicine, and in industry. Illustrated with photographs.

124. Thring, M. W. *Robotics and Telechirs: Manipulators with Memory; Remote Manipulators; Machine Limbs for the Handicapped.* Chichester, England & New York: Ellis Horwood, Ltd. & Halsted Press, 1983. 298 p.
Considers "What does man want from slave machines?"; man as a machine and machine maker; the man/machine interface; sensors for robots; telechirs and aids for the handicapped (sceptres); mechanical arms and hands; walking machines; the current state of theory and practice of robots; telechiric machines and the use of robots; and telechirs and mechanochiropods in the creative society. Illustrated with photographs and drawings.

125. Time-Life Books Staff. *Robotics*. Alexandria, VA: Time-Life Books, 1987. 128 p.
Covers the search for surrogates, robot sight, robot arms, speech, mobility, legged locomotion, dexterity and touch, super skills, automated guided vehicles, and uses in dangerous jobs. Illustrated with color and black-and-white drawings and photographs.

126. Voelker, Alan. "Cluster Review: The Age of Robotics." *Curriculum Review* 24 (January/February 1985), 31.

Gives reviews on three books on robotics suitable for high school students: *The VOR*, by Tom Lonergan and Carl Frederick (see number 695), *Robotics Age: In the Beginning*, by Carl T. Helmers (see number 92), and *Android Design: Practical Approaches for Robot Builders*, by Martin Bradley Weinstein (see number 814).

127. Walters, Wally W. "The Robots Are Coming!" *Journal of Employment Counseling* 19 (March 1982), 45-48.
Discusses books on robots, definitions of varieties of imitation people, laws of robotics, and industrial robots.

128. Warring, Ronald Horace. *Robots & Robotology*. Blue Ridge Summit, PA: TAB Books, Inc., 1983. 128 p.
A basic overview of the field. Includes popular and industrial robots. Covers robot systems, power, mobility, sensors, the robot brain, and applications, as well as other aspects of the robot. Profusely illustrated with drawings. Includes photographs.

129. Webb, Michael. "The Robots Are Here! The Robots Are Here!" *Design Quarterly* (Spring 1983), 4-21.
Covers history, robots in literature, motion pictures, and art, applications, especially in Japan, and the future. Illustrated with photographs and drawings.

130. Wellborn, Stanley N. "Machines That Think: They're Brewing a Revolution." *U. S. News & World Report* 95 (December 5, 1983), 59-60, 62.
Reports on machine intelligence progress, and includes a report on progress in research on robots. Illustrated with color photographs. Cover of this magazine is a color picture of a robot posing as a thinker.

131. Witty, Susan, and Heisler, Gregory. "The Robots Are Coming." *Life* 7 (December 1984), 168-172, 174, 176.
Reports on ways robots are being used. Illustrated with color photographs.

"Picture Books"

132. Gatland, Kenneth, and Jefferis, David. *Robots; The World of the Future; Science & Medicine into the 21st Century*. Illustrated by Terry Hadler, et al. London: Usborne Publishing, Ltd., 1979. 32 p.

A colorful picture book on many types of machines, including robots, that are being used now or will be used in the future. Covers the areas of energy and resources, including mining in space, ecology and agriculture, health and medicine, artificial intelligence, military uses, speaking to the stars, extrasensory perception, and the next 120 years.

133. Henson, Hilary. *Robots*. New York: Warwick Press, 1981. 77 p.

Covers history of robots, science fiction robots, movie robots, robots in industry and the home, robots that "think," the robot brain, and robots and the future. Includes color and black-and-white photographs throughout in picturebook format. Provides glossary.

134. Lambert, Mark. *Robots*. London: Macdonald Educational Ltd., 1981. 32 p.

An all-color, illustrated introduction to robots, their applications, and future use.

135. Malone, Robert. *The Robot Book*. New York: Jove Publications, Inc., 1978. 159 p.

Picture book on robots. Covers what a robot is, origins and history, the robot as hero-villain, robots at work, robots as entertainers, robots in art, inside the robot, how to build a robot, coping with robots, and the future.

136. McKie, Robin. *Robots*. New York: Franklin Watts, 1986. 32 p.

Includes robot anatomy, movement, programming, powering a robot, sensors, vision, robots at work, the future, and mobile, personal, educational, and intelligent robots, and a date chart. Illustrated with color photographs and drawings.

137. McKie, Robin. *The How and Why Wonder Book of Robots*. Los Angeles: Price/Stern/Sloan, 1987. 32 p.
Identical with *Robots*, number 136.

138. Morris, Brian. *The World of Robots*. New York: W. H. Smith Publishers, Inc., 1985. 167 p.
Covers the robot imperative, creating robots,the robot brain, robots in industry, home, space, and science fiction, and the future. Illustrated with color and black-and-white photographs and drawings.

139. Potter, Tony, and Guild, Ivor. *Robotics*. Illustrated by Jeremy Gower. London: Usborne Publishing, Ltd., 1983. 48 p.
A book of color and black-and-white drawings, accompanied by short sections of text. Covers what robots can and cannot do, designing robots, how robots are taught and controlled, sensors, types of robot arms and how they work, cybernetics, mobile, industrial, special purpose, space, home, and computer-controlled robots, how to build a mobile robot, how to make a pneumatic gripper from household articles, and recent developments, and provides a glossary.

140. Vowles, Andrew, and Mackie, Dan. *Robotics*. Illustrated by Charles E. Bastien. Milwaukee, WI: Penworthy Publishing Company, 1986. 32 p.
Covers uses of robots in home, commerce, and industry, how they work, history, robot features, human-machine use, the future, and robotic words. Illustrated with color drawings.

Preschool Level

141. Petty, Kate. *Robots*. Illustrated by Mike Saunders and Adam Wills. New York: Franklin Watts, 1984. 30 p.

Preschool and early grades level. Tells what a robot is, what robots can do, how some types are controlled, and how robots are used. Includes review questions. Illustrated with color pictures.

Elementary Grades Level

142. "Electronic Robots Designed to Help Out Humans." *Instructor* 93 (December 1983), 84.
Reports on uses of robots and the origin of the word "robot."

143. "Here Come the Robots!" *National Geographic World* no. 138 (February 1987), 18-21.
Describes what robots can do, and reports on various applications. Illustrated with color photographs showing a robotic cat, toy robots, a robot serving in a restaurant, a robot family, and a ComRo robot feeding a seal.

144. "Robots! Is There One in Your Future?" *National Geographic World* 37 (September 1978), 22-29.
Describes robot uses now and in the future, and robot designers, and includes instructions and design for a do-it-yourself robot tee shirt. Illustrated with color photographs.

145. "RRRR...RRRRR...Robots!" *National Geographic World* no. 109 (September 1984), 4-9.
Discusses ways robots are being used. Illustrated with color photographs. Cover of this issue features Orlando (Florida) Police Department's educational robot.

146. Asimov, Isaac. *How Did We Find Out About Robots?* Illustrated by David Wool. New York: Walker and Company, 1984. 62 p.
Covers history and legends of robots, Asimov's role in the story of robots, science fiction and robots, industrial robots,

robots and people, and robots of the future. Illustrated with drawings.

147. Barrett, N. S. *Robots*. Illustrated by Mike Saunders and Stuart Willard. New York: Franklin Watts, Inc., 1985. 32 p.

Describes what a robot is and how it works, and gives applications in work, dangerous jobs, space, education, and for play. Also has a short history, facts and records, and glossary. Includes color photographs and drawings.

148. Berger, Melvin. *Robots in Fact and Fiction*. New York: Franklin Watts, 1980. 86 p.

Covers robots in history, mechanical and electrical robots, robots at work, robots of myth and magic, human robots, and science fiction robots. Includes photographs and drawings. Suitable for junior and senior high levels.

149. Billard, Mary. *All About Robots*. Illustrated by Walter Wright. New York: Grosset & Dunlap, 1982. 45 p.

Describes history of robots, what robots are, robots around the house, in industry, in research, in space, and in fantasy and film, and robots of the future. Illustrated with color photographs and drawings.

150. Darling, David J. *Robots and the Intelligent Computer*. Illustrated by Tom Lund. Minneapolis, MN: Dillon Press, Inc., 1986. 77 p.

First section of book deals with computers. Second section discusses robots. Covers how a robot is made, robots in space, the factory, and home, and what robots will do in the future. Provides a section on possible subjects for classroom discussion or debate concerning computers and robots. Includes glossary. Illustrated with color photographs and drawings.

151. Davis, Charles. "Meet the Robots." *Child Life* 65 (June/July 1986), 4-9.

Discusses history of robots and what robots can do. Includes an exercise page for reader to design his or her own robot, along with sample design. Illustrated with drawings and photographs. Cover of this issue shows a group of robots.

152. Elting, Mary, and Folsom, Michael. *The Answer Book About Robots and Other Inventions*. Illustrated by Rowan Barnes-Murphy. New York: Grosset & Dunlap, 1984. 80 p.
A book in question-and-answer format that covers information on what a robot is, what it can do, how some of its senses and mechanisms work, and related topics. Illustrated with drawings.

153. Greene, Carol. *Robots*. Chicago: Childrens Press, 1983. 48 p.
Discusses what a robot is, history of robots, what robots can do, why they are used, and how they are used. Includes a glossary. Illustrated with color and black-and-white photographs.

154. Hawkes, Nigel. *Robots and Computers*. New York: Franklin Watts, 1984. 29 p.
Covers what a robot is, science fiction robots, industrial robots and how they work, programming a robot, remote control of robots, robots in transportation, war, and space, and robots of the future. Includes glossary. Illustrated with color photographs and drawings.

155. Jefferis, David. *Robots*. Illustrated by Christopher Forsey, the Hayward Art Group, and Michael Roffe. New York: Franklin Watts, Inc., 1982. 32 p.
Covers what a robot is, its brain and senses, and applications. Illustrated with color drawings and photographs.

156. Kelly, Derek. *Ollie the Robot...Talks About Robots*. Illustrated by Christine Fedorowicz. New York: Petrocelli Books, Inc., 1987. 121 p.
Covers what a robot is, robot anatomy, how robots work, how a robot looks, and robots in the world and universe. Illustrated with drawings.

157. Kerker, Alexander. *Robots*. Illustrated by Tom LaPadula. New York: Western Publishing Company, Inc., 1984. 45 p.

Examines what robots are, history of robots, applications, and robots of the future. Illustrated with color drawings.

158. Kerrod, Robin. *Robots*. Illustrated by Tudor Art Studios. Vero Beach, FL: The Rourke Corporation, 1984. 64 p.

Covers what a robot is, history of robots, robots in fiction, what robots do, how they are controlled, applications, and robots in space. Illustrated with color pictures.

159. Kleiner, Art. *Robots*. Illustrated by Jerry Scott. Milwaukee, WI: Raintree Publishers, 1981. 48 p.

Covers what a robot is, history of robots, kinds, inside a robot, the robot's mind and senses, teaching a robot, and the future with robots. Includes glossary. Illustrated with color drawings.

160. Knight, David C. *Robotics; Past, Present, & Future*. New York: William Morrow & Company, 1983. 122 p.

Covers what a robot is, history, applications in industry, home, office, science, medicine, and space, and the future. Includes photographs and drawings.

161. Lambert, Mark. *50 Facts About Robots*. New York: Warwick Press, 1983. 32 p.

Answers 50 questions, each illustrated with color drawings, concerning what robots are, how they work, how they are used, and how they may be used in the future.

162. Lambert, Mark. *Just Look At...Robotics*. Vero Beach, FL: Rourke Enterprises, Inc., 1985. 45 p.

Covers what a robot is, robots in use, robot senses and movement, and robots in the future. Includes chronology. Illustrated with color photographs and drawings.

163. Lauber, Patricia. *Get Ready for Robots*. Illustrated by True Kelley. New York: Thomas Y. Crowell, 1987. 32 p.

Tells about personal, industrial, underwater, and other types of robots, how they see and move, and what robots may do in the future. Illustrated with color drawings.

164. Lindblom, Steven. *How to Build a Robot*. New York: Thomas Y. Crowell, 1985. 78 p.
Discusses what a robot is, history of robots, and what makes up a robot. Covers motion, sense, thought, and overall makeup of a robot, using the building of a hobby robot as illustration. Also includes information on applications. Illustrated with photographs and drawings.

165. Liptak, Karen. *Robotics Basics*. Illustrated by Mike Petronella. Englewood Cliffs, NJ: Prentice-Hall, Inc., 1984. 48 p.
Covers the origin of robots, what a robot is, why they are used, how robots are used, how they are controlled, and future uses. Includes a glossary. Illustrated with photographs and drawings.

166. Litterick, Ian. *Robots and Intelligent Machines*. Illustrated by David Anstey. New York: The Bookwright Press, 1984. 47 p.
Discusses what robots are, their history, how they are used, robots in space, the morals of robots, and the difference between robots and human beings. Includes glossary and color illustrations.

167. Maccarone, Grace; Krulik, Nancy; and Epstein, Jolie. *Real Robots*. New York: Scholastic Inc., 1985. 32 p.
Shows color photographs of robots and their applications, explained in text accompanying the pictures.

168. Marrs, Wanda, and Marrs, Texe. *The Great Robot Book*. New York: Julian Messner, 1985. 104 p.
Covers what robots are, types of robots, history of robots, how robots work, how robots are used, robots as pets and pals, toy robots, robot kits for making robots, robots in space, robots in the movies, robot facts and trivia, and how to learn more

about them. Illustrated with photographs and drawings. Describes several educationally useful robots.

169. Marsh, Peter. *Robots*. New York: Warwick Press, 1983. 37 p.

Discusses what a robot is, what an industrial robot is, how they are taught and controlled, how robots are used in industry, space, and other areas, robots in fiction, robots built by robots, and the robot future. Includes glossary. Illustrated with color photographs and drawings.

170. Metos, Thomas H. *Robots A2Z*. New York: Julian Messner, 1980. 80 p.

Discusses what a robot is, history of robots, industrial robots, robots in movies and television, robots in space and war, robots around the house, and robots of the future. Illustrated with drawings and photographs.

171. Milton, Joyce. *Here Come the Robots*. Illustrated by Peter Stern. New York: Hastings House, 1981. 118 p.

Covers history of robots, applications such as in homes, business, industry, and outer space, cyborgs, and the future. Illustrated with photographs and drawings.

172. Paltrowitz, Stuart, and Paltrowitz, Donna. *Robotics*. New York: Julian Messner, 1983. 63 p.

Discusses how robots work, personal robots, industrial robots, office robots, androids, famous robots on film, and robots of the future. Illustrated with photographs.

173. Rickard, Graham. *Robots*. East Sussex, England: Wayland (Publishers) Ltd., and New York: The Bookright Press, 1986. 32 p.

Covers what a robot is, robots at work, the next robots, and robots of the future. Illustrated with color photographs and color and black-and-white drawings..

174. Rutland, Jonathan. *The World of Robots*. New York: Warwick Press, 1979. 23 p.

Covers early robots, current applications, factory robots, robots in research, mobile "wanderer" robots, robot warriors, robot brains, the use of robots to extend human capabilities, and robots of the future. Illustrated with color drawings.

175. Ryder, Joanne. *Star Wars; C-3PO's Book About Robots*. Illustrated by John Gampert. New York: Random House, 1983. 30 leaves.
Covers history of robots, entertainment robots, industrial robots, robots underseas, in space, in the home, and other applications, how robots work, and the future. Illustrated with color drawings.

176. Storrs, Graham. *The Robot Age*. New York: The Bookwright Press, 1985. 48 p.
Covers history of robots, what a robot is, how robots learn, robot senses, working robots, robots working in dangerous places, and personal robots. Illustrated with color photographs and drawings.

Secondary Grades Level

177. Baldwin, Margaret, and Pack, Gary. *Robots and Robotics*. New York: Franklin Watts, 1984. 61 p.
Provides information on the history of robots, the robotic mind and body, industrial robots, robots in space, robots at home, the possibility of robots with emotions in the future, and career information. Includes a robot-related, computer programming activity. Illustrated with photographs and drawings.

178. Bendick, Jeanne. *Super People; Who Will They Be?* New York: McGraw-Hill Book Company, 1980. 128 p.
Includes discussions on clones, bionic people (humans with substitute parts), androids, and robots. Illustrated with photographs and drawings.

179. Chester, Michael. *Robots; Facts Behind the Fiction*. New York: Macmillan Publishing Company, 1983. 90 p.

Surveys various types of robots, turtle robots, robot arms, and applications in space, and robots as sentries, for fun, at work, and in the future. Includes encyclopedic glossary. Illustrated with photographs.

180. D'Ignazio, Fred. *Working Robots*. New York: E. P. Dutton, 1982. 149 p.

Provides information on the working robot. Covers types and parts of robots, how to teach a robot, robots and the outside world, robots in the home, in industry, in exploration, and in the classroom, and the future. Also gives information on building one's own robot. Has a glossary and list of related materials. Includes photographs.

181. Geil, John J. "Robotics." *Industrial Education* 72 (January 1983), 51.

Presents a quiz on industrial robots, with answers.

182. Osborne, David M. *Robots; An Introduction to Basic Concepts and Applications*. Detroit, MI: Midwest Sci-Tech Publishers, Inc., 1983. 210 p.

Primarily on industrial robots, but includes information on other applications. From the industrial robot standpoint, covers what a robot is, types of robots, what makes up a robot, robot sensors, how a robot works, how it is controlled, robot maintenance, types of jobs for which robots are suited, and future uses. Illustrated with drawings.

183. Pawson, Richard. *The Robot Book*. London: Frances Lincoln, Ltd., 1985. 192 p.

Covers history of robots, robots in action, how robots work, and practical projects - robot devices to build and control with a home computer. Illustrated with color and black-and-white photographs and drawings.

184. Salant, Michael A. *Introduction to Robotics*. New York: McGraw-Hill, 1988. 121 p.

Covers introduction to robots, makers of robots, users of robots, and robotics researchers, and includes student review questions and additional sources of information. Illustrated with photographs and drawings.

185. Sharp, Mike. *Robot World.* New York: Warwick Press, 1984. 93 p.

Covers what robots are, history of robots, laws of robots, robot senses, how robots move, the robot brain, whether robots can think, artificial intelligence research, and applications of robots in the workplace, at home, and in the future. Includes glossary. Illustrated with color photographs and drawings.

186. Silverstein, Alvin, and Silverstein, Virginia B. *The Robots Are Here*. Englewood Cliffs, NJ: Prentice-Hall, Inc., 1983. 128 p.

Deals with what a robot is, applications, building one's own robots, robots for fun, and robots versus humans. Includes photographs.

187. Winter, Ruth. "Heady Metal." *Seventeen* 42 (August 1983), 224.

Reports on ways robots are used in the home and in business.

ROBOT FEATURES

Arms

188. "A Polyurethane Proboscis." *Scientific American* 256 (April 1987), 73-74.

Reports on a two-foot-long, pneumatic, polyurethane trunk, for use as a flexible robot arm, built by a team led by James F. Wilson of Duke University, Durham, North Carolina. Includes color photograph.

189. "Compact Robot Moves in Seven Axes." *Machine Design* 56 (June 7, 1984), 83.

Describes robot arm with ability to get into cramped places inaccessible to a conventional robot, designed by Spine Robotics, Molndal, Sweden. Illustrated with photograph and drawings.

190. "Mathematics Frees Robots for Finer Tasks." *New Scientist* 119 (August 25, 1988), 35.

Reports on research on robot arm movement being conducted by Eduardo Bayo at the University of California in Santa Barbara. Illustrated with photograph.

191. Brown, Stuart F. "Rubber-Armed Robot." *Popular Science* 226 (May 1985), 28.

Reports on an industrial-type robot with rubber "muscles" and a rubber "arm," developed by Bridgestone Corporation and Hitachi, Limited. Includes photograph and drawing.

192. Capulong, Eduardo R. C. "Baseball Is for Robots." *Popular Science* 231 (August 1987), 39.

Reports on research by Ron Huston and Tim King of the University of Cincinnati, Ohio, to analyze the baseball pitch, to aid in building less jerky and more energy efficient robots.

193. Hawker, J. Scott; Nagel, R. N.; Odrey, Nicholas G.; and Roberts, Richard. "Multiple Robotic Manipulators." *Byte* 11 (January 1986), 203-204, 206, 208, 211-212, 214, 216, 219.

Discusses the use and problems of a dual-arm robot, and coordinating two robots, in integrating robots to work together on the same task or in the same workplace. Illustrated with drawings.

194. Mortimer, John. "Personal Robots Set to Put Glamour Back into Robotics." *The Industrial Robot* 14 (June 1987), 77-79.

Describes robotic arm designed for light work such as health care, produced by Universal Machine Intelligence in England. Illustrated with photographs.

195. Singer, Andrew, and Rony, Peter. "Controlling Robots with Personal Computers." *Machine Design* 54 (September 23, 1982), 78-82.

Discusses use of small industrial robot arms that can perform light production tasks. Illustrated with photographs and drawings.

196. Solomon, Leslie. "Robotic Arms." *Computers & Electronics* 21 (February 1983), 74-75.

Describes types of robotic arms. Includes drawings.

Autonomous Land Vehicles and Navigation

197. "Lab on Wheels Tests AI." *Design News* 44 (May 23, 1988), 24.

Reports on the Autonomous Land Vehicle, a driverless vehicle designed to test a computer's ability to "think" on the go, developed by Martin Marietta Corporation, Denver, Colorado. Illustrated with color photograph.

198. "Robots Hit the Road." *Popular Mechanics* 165 (November 1988), 16.

Reports on United States Army's Soldier-Robot Interface Project robot utility cart, developed at Oak Ridge National Laboratories, Tennessee, and NavLab (navigation laboratory), a Chevrolet van that drives itself, being developed at Carnegie Mellon University, Pittsburgh, Pennsylvania. Illustrated with color photographs.

199. "Self-Steering Robot." *Popular Mechanics* 167 (January 1990), 16.

Reports on robot that can steer itself and avoid obstacles at 1.5 miles per hour, developed by Mazda in Yokohama, Japan. Illustrated with color photograph.

200. Fisher, Arthur. "Mobile Robot - I; Mobile Robot - II." *Popular Science* 231 (December 1987), 12.

"Mobile Robot - I" reports on free-ranging robots called automatic guided vehicles (AGVs) that can find their way around a factory floor, developed by Colin Besant and colleagues at Imperial College of Science and Technology, London, England. "Mobile Robot - II" reports on work at Sandia National Laboratories on a robot for sentry duty. Illustrated with color photographs.

201. Jorgensen, Charles; Hamel, William; and Weisbin, Charles. "Autonomous Robot Navigation." *Byte* 11 (January 1986), 223-224, 226, 228, 230-232, 235.

Discusses self-location and navigation by autonomous robots and ways they can find the best path to a goal or destination. Illustrated with photograph and drawings.

202. Schefter, Jim. "Look Ma! No Driver." *Popular Science* 227 (October 1985), 94-95, 114-115.

Describes the Autonomous Land Vehicle developed by Martin Marietta Aerospace, Denver, Colorado. Illustrated with photograph and drawings.

Computer Program Languages

203. "Researchers Work Toward One Software Language for Robots." *Research & Development* 27 (August 1985), 56-58.

Reports on work at University of Southern California, Los Angeles, to develop a uniform computer programming language for robots.

204. Blume, Christian, and Jakob, W. *PASRO: Pascal for Robots*. Berlin: Springer-Verlag, 1985. 133 p.

Describes a programming system for teaching, studying, and experimenting. Intended for robotic arms.

205. Blume, Christian; Jakob, W.; and Favaro, J. *PasRo; Pascal and C for Robots*. 2nd ed. Berlin: Springer-Verlag, 1987. 239 p.

Provides introduction to robot programming, description of PASRO system concept and Pascal and C languages, and reference information on PASRO and PASRO/C.

206. Bonner, Susan, and Shin, Kang G. "A Comparative Study of Robot Languages." *Computer* 15 (December 1982), 82-96.
Classifies and compares 14 computer-programming robot languages used for computer control of industrial robots.

207. Rony, Peter R., and Rony, Karl E. *Introduction to Robot Programming in BASIC*. Illustrated by Paul A. Rony. Reston, VA: Reston Publishing Company, Inc., 1985. 376 p.
Covers what a robot is, review of CP/M, hierarchies, architectures, and Microsoft BASIC, Mitsubishi RM/101 Movemaster Industrial Robot, Movemaster robot command language, program mode of robot operation, and introduction to three-dimensional robot joint motions. Illustrated with drawings.

208. Storrs, Graham. "A Sense of Awareness: Robots Have a Long Way to Go Before They Conquer the World." *P C World* 9 (January 1986), 168-174.
Discusses ways of controlling robots through computer programming languages. Includes illustrations.

209. Tomek, Ivan. "Josef, the Robot." *Computers and Education* 6 (1982), 287-293.
Describes Josef, a robot programming language similar to Logo.

Hands and End Effectors

210. "Fingers Help Robot Hand." *Popular Mechanics* 165 (September 1988), 18.
Reports on robot hand designed by Uri Tsach at Pennsylvania State University, University Park. Illustrated with color photograph.

211. "Gripping Device Provides Robots with 'Sense of Touch.'" *Research & Development* 27 (August 1985), 58, 60.

Describes robot gripper developed at Imperial College, London, England. Illustrated with photograph.

212. "Hand Holder." *Popular Science* 235 (December 1989), 72.

Reports on handworn apparatus for teaching robot hand to perform task. Illustrated with color photograph.

213. Kato, Ichiro. *Mechanical Hands Illustrated*. Tokyo: Survey Japan, 1982. 214 p. Available from Hemisphere Publishing Corporation, 79 Madison Avenue, Suite 1110, New York, NY.

An English reprinting of the original 1977 Japanese edition. Covers all types of artificial and mechanical hands for prosthetic, industrial, and other uses. Provides photographs, descriptions, and drawings of over 100 different models. Also provides drawings and information on over 100 other models that have been patented.

214. Kato, Ichiro, and Sadamoto, Kuni, eds. *Mechanical Hands Illustrated*. Rev. ed. New York: Hemisphere Publishing Corporation, 1987. 216 p.

Contributions cover current trends in research and development, mechanism of artificial hands, analysis of operations, grasping in handwork, and power sources. Also includes section on patents and utility models. Provides photographs, drawings, and information on many mechanical hands available.

215. Kinnucan, Paul. "Robot Hand Approaches Human Dexterity." *High Technology* 5 (March 1985), 80.

Describes robot gripper developed by Hitachi in Japan. Illustrated with photograph.

216. Mason, Matthew, and Salisbury, J. Kenneth, Jr. *Robot Hands and the Mechanics of Manipulation*. Cambridge, MA: The MIT Press, 1985. 298 p.

Technical coverage of robot hands and manipulation. Illustrated with photographs and drawings.

217. Mishara, Eric. "Omelet Robot." *Omni* 8 (February 1986), 32.
Reports on dexterous robotic hand being developed by Stephen Jacobsen at the University of Utah, Salt Lake City. Illustrated with photograph.

218. Ottinger, Lester V. "Design and Use Considerations for End-of-Arm Tools." *Industrial Engineering* 14 (October 1982), 24-26, 28, 30, 32.
Discusses end-of-arm tools for industrial robots.

219. Smith, Emily T. "Short-Handed in the Lab? This Robot Could Be the Answer." *Business Week* no. 2686 (February 23, 1987), 122.
Reports on Zymark Corporation robot programmed to use four specialized hands, by researchers at Carnegie Mellon University, Pittsburgh, Pennsylvania.

220. Stambler, Irwin. "Research on Mechanical Hands Relies on Innovative Software." *Research & Development* 28 (January 1986), 48-49.
Reports on research to develop a robot hand with sensors on each fingertip.

Hearing

221. Siwolop, Sana. "Now Robots Are Getting Hearing Aids." *Business Week* no. 3001 (June 1, 1987), 95.
Reports on small microchip that detects sound, to be used by robots. Includes color drawing.

Intelligence

222. "'Heuristic Robots Could Boost Slumping U. S. Productivity.'" *Research & Development* 29 (May 1987), 58, 61.

Discusses possibility of developing heuristic capability in robotic systems to give a robot judgment using the following steps: doing what it is told to do, observing what takes place, and then doing another operation as a result of its action.

223. Chafetz, Michael. "Will Power." *Science Digest* 91 (April 1983), 75.
Reports on Cerebellar Model Arithmetic Computer (CMAC), developed by Dr. James Albus, National Bureau of Standards, that could be used with a robot to give it a human brain mimic. Cover of this issue shows color drawing of robot.

224. Cole, Tim. "Thinking Robots." *Popular Mechanics* 166 (June 1989), 21.
Reports on research on neural robots that use neural network (brain-mimicking) computers to increase robot intelligence, by Dr. Michael Kuperstein of Neurogen Laboratories, Brookline, Massachusetts.

225. Fleet, Kenneth. "A Robot Era?" *World Press Review* 32 (January 1985), 37-38.
Discusses needs for industrial robots to see, hear, and infer.

226. Lee, Mark H. *Intelligent Robots*. New York: Halsted Press, 1989. 210 p.
Discusses problems and issues to consider when designing intelligent factory automation systems, particularly those using robots. Illustrated with drawings.

Legs

227. "Putting on a Robotic Balancing Act." *Science News* 132 (July 25, 1987), 63.
Reports on a six-foot-tall robot able to balance itself on one leg, developed by Irving J. Oppenheim and Lyman

Petrosky at Carnegie Mellon University in Pittsburgh, Pennsylvania.

228. "Robot Biped." *Popular Mechanics* 164 (August 1987), 96.
Reports on biped robot that walks, developed by Dr. Yuan Zheng, Clemson University, Clemson, South Carolina. Illustrated with color photograph.

229. "Robot Leg Learns to Hop." *Science Digest* 8 (September 1981), 26.
Briefly describes the single hopping aluminum leg being developed by Edward Fredkin and associates at Massachusetts Institute of Technology, Cambridge. Includes photograph and sketch of hop.

230. "Robotics Take a Natural Turn." *New Scientist* 114 (May 21, 1987), 36.
Reports on study of leg motions and walking behavior of animals such as insects, spiders, crabs, and lobsters to help in developing walking machines, being conducted at Oregon State University, Corvallis. Illustrated with photograph.

231. "Robots Mimic Spider's Step." *Science Digest* 90 (December 1982), 23.
Describes research studying daddy-longlegs at Ohio State University to aid in the design of six-legged walking robots. Illustrated with drawings.

232. "Walk This Way..." *Popular Mechanics* 162 (April 1985), 68-69.
Reports on walking robots. Includes color photographs of four examples.

233. Britton, Peter. "Engineering the New Breed of Walking Machines." *Popular Science* 225 (September 1984), 66-69.
Reports on progress in developing walking robots. Discusses Odex I (Odetics, Incorporated, Anaheim, California), based on a personal demonstration of its capabilities given to the author. Includes photographs.

234. Peterson, Ivars. "See How They Run: Motion Symmetry." *Science News* 29 (March 15, 1986), 167.
Reports on research on running conducted by robot designer Marc H. Raibert at Carnegie Mellon University, Pittsburgh, Pennsylvania.

235. Port, Otis. "Want to Design a Robot? Try Watching a Bug." *Business Week* no. 2989 (March 16, 1987), 101.
Reports on research studying motion of insects and spiders to determine how to create walking robots. Includes color drawing.

236. Raibert, Marc H. *Legged Robots That Balance*. Cambridge, MA: The MIT Press, 1986. 233 p.
Provides a technical introduction to the study of legged machines, including a history of research in the field, hopping on one leg in the plane, hopping in three dimensions, biped and quadruped running, symmetry in running, alternatives for locomotion control, tabular control of running, and research on animals and vehicles. Illustrated with photographs and drawings.

237. Raibert, Marc H. "Symmetry in Running." *Science* 231 (March 14, 1986), 1292-1294.
Discusses the role of symmetry in simplifying the control of legged robots and in giving them the ability to run and balance. Also considers studies of applicable symmetrics. Includes photographs and drawings.

238. Raibert, Marc H., and Sutherland, Ivan E. "Machines That Walk." *Scientific American* 248 (January 1983), 44-53.
Describes the work of the authors with machines (robots) that hop, crawl, walk, and run. Discusses the problems of developing such machines. Provides drawings of a six-legged machine that crawls, and hopping machines. Cover of this issue shows a color drawing of a crawling machine.

239. Todd, D. J. *Walking Machines; An Introduction to Legged Robots*. London: Kogan Page, 1985. 190 p.

Provides an introduction to the study of legged machines, including a history of walking machines, general principles of legged locomotion, mechanics of legged vehicles, analysis and control, computing requirements, current developments, and applications of walking machines. Illustrated with photographs and drawings.

240. Wallach, Leah. "Walking Machine." *Omni* 8 (January 1986), 30.

Describes a computer-controlled, six-legged, adaptive suspension vehicle developed at Ohio State University, Columbus. Illustrated with photograph.

Microrobots

241. "Microscopic Robots in Development." *High Technology Business* 9 (February 1989), 37.

Reports on efforts at the University of California at Berkeley to develop microscopic structures to assemble tiny robots.

242. Gannon, Robert. "Micromachine Magic." *Popular Science* 234 (March 1989), 88-92, 143.

Reports on research on microscopic motors and robots, and explains how they work. Illustrated with color and black-and-white photographs and drawings.

243. Heppenheimer, T. A. "Microrobots." *Discover* 10 March 1989), 78-82, 84.

Reports on research to develop microscopic machines and robots. Illustrated with color photographs and drawings.

244. Hoffer, William. "Micromachines." *Popular Mechanics* 165 (July 1988), 79-81, 121.

Discusses micromachines, including microrobots (gnat robots), and ways they may be used. Illustrated with color drawings and photographs.

245. McDermott, Jeanne. "The Coming Age of Micromachines." *Popular Science* 224 (June 1984), 87-91.
Discusses the field of micromechanics, including microrobots, and research being done. Illustrated with color photographs and drawings.

Miscellaneous

246. "Hypercube Robot." *Popular Mechanics* 167 (June 1990), 20.
Reports on HERMES III, a robot with parallel processing super computers being developed at Oak Ridge National Laboratories, Oak Ridge, Tennessee. Illustrated with photograph.

247. "Massive Robot Will Test Robotic Theories." *Research & Development* 28 (January 1986), 41.
Reports on the Adaptive Sensing Vehicle, a six-legged robot being developed by Ohio State University, Columbus.

248. "Metallic Recall." *Popular Science* 233 (December 1988), 72.
Reports on Space Wings, a robot toy built using space-memory alloys. When heated, wire contracts and flaps the toy's wings. As it cools, the wire expands and the wings move in the opposite direction. Illustrated with color photograph. See number 249.

249. "Mondo-Tronics Space Wings Robotics Kit." *Radio-Electronics* 58 (October 1987), 22-23.
Describes a robotics kit that builds a pair of flapping wings, available from Mondo-Tronics (20090 Rodriques Avenue #1, Cupertino, California 95014), which uses Bio-Metal wire, a shape-memory alloy wire.

250. "Robot Turns into a Work-Mate." *New Scientist* 98 (April 28, 1983), 223.

Describes Yes-Man, a two-armed robot designed to be used with human workers in any sequence of tasks that requires testing before each step.

251. "The Amazing Micromouse." *The Futurist* 13 (December 1979), 481-482.
Reports on Moonlight Special, a robot mouse that can find its way through a maze, and ways robots of this type could be used in the future. Illustrated with photographs.

252. Albus, James S., and Evans, John M., Jr. "Robot Systems." *Scientific American* 234 (February 1987), 76-86B.
Discusses feedback and control systems for industrial robot control, and robotic sensing. Illustrated with photographs and drawings.

253. Darrow, Barbara. "In Quest of the 'Humanoid' Robot." *Design News* 44 (March 21, 1988), 48-52, 54.
Reports on research to give robots senses of sight, smell, hearing, touch, and taste, and the ability to learn from experience. Illustrated with color and black-and-white photographs and drawing.

254. Dworetzky, Tom. "Dextor, the Willowy Six-Footer That Bends at the Waist." *Discover* 8 (October 1987), 18.
Reports on a robot able to balance on two wheels while lifting and carrying half its weight, developed by Lyman Petrosky at Carnegie Mellon University in Pittsburgh, Pennsylvania. Includes color photograph.

255. Finkel, James I. "Computers and Education." *School Product News* (May 1985), 8.
Discusses mechanical dynamics problems of mobile robots and robot arms.

256. Fisher, Arthur. "Humanoid? Android? Robot?" *Popular Science* 227 (July 1985), 8.
Describes prototype anthropomorphic robot, McAndroid the Android, invented by Joe Barron of McAndroids, Limited,

in England, to use as a test-bed for new robot technology. Illustrated with photograph.

257. Fisher, Arthur. "Robot I: Gadfly; Robot II: Omnigripper." *Popular Science* 227 (November 1985), 16.
"Robot I: Gadfly" describes a robot with three pairs of telescoping legs, developed by GEC Laboratories, Great Baddow, Essex, England. "Robot II: Omnigripper" describes a robot gripper invented by Peter Scott at Imperial College, London, England. Includes photographs.

258. Hafner, Katherine. "Robotics Research Center Logs Decade Mark." *Computerworld* 17 (April 4, 1983), 34.
Describes the Robotics Research Center at the University of Rhode Island, Kingston, Rhode Island, which does research on vision, sensing, and integrated robotics systems. Illustrated with photograph.

259. Hapgood, Fred. "Building a Better Mouse." *Omni* 12 (November 1989), 22, 126.
Describes work of Dave Otten with maze-solving robot mice at Massachusetts Institute of Technology, Cambridge. Illustrated with photograph.

260. Hapgood, Fred. "MIT's Roach Motel." *Omni* 11 (October 1988), 38, 173.
Describes robots that move like insects, being developed by Rodney Brooks at Massachusetts Institute of Technology, Cambridge. Illustrated with photographs.

261. King, Jean Progner. "Robots Will Never Be Practical Unless Products Are Designed for Them." *Industrial Design* 29 (January/February 1982), 24-29.
Discusses parts design and the use of robots in industry, and provides rules of thumb on how to design for robots. Illustrated with photographs.

262. Lawren, Bill. "Recipes for Robots." *Omni* 8 (September 1985), 22.

Describes research on use of robots for demand production of parts, which would enable factories to produce parts from the past.

263. Lawren, Bill. "Rollerbot." *Omni* 11 (December 1987), 44.
Reports on roller-skating robot that can keep its balance, invented by Irving Oppenheim and Lyman Petrosky at Carnegie Mellon University, Pittsburgh, Pennsylvania. Illustrated with photograph.

264. Peterson, Ivars. "Robots with a Lot of Nerve." *Science News* 131 (June 6, 1987), 362-363.
Reports on work of Michael Kuperstein at Wellesley College, Massachusetts, on "eye-hand" coordination for a robot. Includes drawings. Cover of this issue shows a robot.

265. Scibor-Rylski, M. "'Yes-Man' - A Co-Operative Robot Workstation." In *Proceedings of the 1st International Conference on Human Factors in Manufacturing, 3-5 April 1984, London, UK*, pp. 41-54. Edited by Tom Lupton. Bedford, England: IFS (Publications), Ltd, 1984. 374 p.
Describes robot designed to work together with human in manufacturing and assembly operations. Illustrated with photographs and drawings.

266. Stover, Dawn. "Coming: Graceful Robots?" *Popular Science* 230 (January 1987), 27.
Reports on industrial robot use of compact motors driven by ultrasonic vibrations which provide smooth, easily controlled rotary motions by the robots. Illustrated with photograph and drawing.

267. Watts, Susan. "Britain Barks at the Heels of Robotics Leaders." *New Scientist* 118 (June 30, 1988), 48.
Reports on the opening of Great Britain's Advanced Robotics Research Centre, designed to pull together research projects on robotics.

268. Weber, David M. "Robots Come to Their Senses." *Electronics Week* 58 (February 18, 1985), 22-24.
Discusses various sensing devices for robots.

269. Weintraub, Pamela. "Raising the Robot's I. Q." *Discover* 2 (June 1981), 77-80.
Describes research to give robots senses of vision, touch, and hearing, and the ability to walk, make decisions, and reproduce themselves. Illustrated with photographs and drawing.

270. Wrigley, Al. "Prototype Expert System to Diagnose, Trouble-Shoot Robots Made by Ford." *American Metal Market* 94 (April 21, 1986), 9-10.
Describes first expert system designed to diagnose and repair robots, using artificial intelligence, developed by engineers at Ford Motor Company's Robotics and Automation Applications Consulting Center, Dearborn, Michigan.

271. Yulsman, Tom. "Can Bioengineering Build a Better Robot?" *Science Digest* 93 (July 1985), 34.
Reports on possible uses of biotechnology to build robots. Illustrated with color photograph.

Smell

272. "Robots That Will Sniff Chemicals." *Chemical Week* 134 (March 14, 1984), 29-30.
Reports on research to develop robots that recognize and remember chemical smells.

273. Mishara, Eric. "Robot Nose." *Omni* 8 (March 1985), 38.
Describes robotic nose based on human nose, being developed at Carnegie Mellon University, Pittsburgh, Pennsylvania. Illustrated with photograph.

Teleoperators and Telepresence

274. MaHill, John. "From Robots to Telepersons." *Technology Review* 83 (January 1981), 82-83.

Discusses the views of Marvin Minsky of the Massachusetts Institute of Technology's Artificial Intelligence Laboratory, Cambridge, concerning telepersons, remotely operated robots that can feel and work like human hands, and "human report interfaces" - the problems of translating human motions to a machine and a machine's senses to a human. Illustrated with drawing.

275. Martin, H. Lee, and Kuban, Daniel P., eds. *Teleoperated Robotics in Hostile Environments*. Dearborn, MI: Society of Manufacturing Engineers, 1985. 273 p.

Includes papers providing an overview of remote manipulation of robots, nuclear applications, undersea applications, space applications, robotic adaptations to hostile environments, and mobile systems. Includes photographs and drawings.

276. Uttal, William R. "Teleoperators." *Scientific American* 261 (December 1989), 124-129.

Discusses teleoperator systems of all types. Illustrated with drawings and color photographs.

277. Vertut, Jean, and Coiffet, Philippe. *Teleoperation and Robotics; Evolution and Development*. Englewood Cliffs, NJ: Prentice-Hall, Inc., 1986. 332 p.

Describes operation of a robot from a distance (teleoperation and telemanipulation). Covers background, system classification, types of manipulators, transporters, vehicles for various applications, such as handling radioactive materials, and other applications. Includes drawings and photographs.

Touch

278. "Getting the 'Feel' of Things." *Design News* 43 (March 23, 1987), 62.

Reports on a synthetic "skin" that could help robots differentiate between surfaces and textures, being developed at the University of Florida at Gainesville's Department of Engineering Sciences.

279. "Gripping Device Provides Robots with 'Sense of Touch.'" *Research & Development* 27 (August 1985), 58-59.

Reports on development of a gripper for robots at Imperial College, London, England. Illustrated with photograph.

280. Edson, Daniel V. "Giving Robot Hands a Human Touch." *High Technology* 5 (September 1986), 31-33, 35.

Examines the progress made on the design of robot fingers. A contact and references to the article are listed on page 69 of this issue. Includes photographs.

281. Evans, Michael R. "Optical Sensor Gives Robots a Light Touch." *Popular Science* 227 (August 1985), 36.

Describes device, a membrane-covered glass, that can give a robot a highly accurate sense of touch, designed by Steve Begej, a graduate student at the University of Massachusetts. Illustrated with photographs and drawing.

282. Hapgood, Fred. "Inside a Robotics Lab: The Quest for Automatic Touch." *Technology Illustrated* 3 (April 1983), 18-22.

Discusses touch sensing robots and devices being developed by John Purbrick at the Massachusetts Institute of Technology, Cambridge. Illustrated with photographs.

283. Lawren, Bill. "Robot Braille." *Omni* 8 (June 1986), 39.

Reports on a silicon and rubber "skin" finger that picks up vibrations as it slides along a surface, for use by robots, developed by Gale Nevill and Robert Patterson at the University of Florida, Gainesville. Illustrated with photograph.

284. Love, Martin. "The Great Robot Fizzle." *Forbes* 136 (November 4, 1985), 226, 228.
Reports on tactile sensing devices being marketed for robot hands. Includes color photographs.

285. MaHill, John. "Giving Robots Touch." *Technology Review* 86 (April 1983), 73.
Describes research to give robots fingers and a sense of touch. Illustrated with drawings.

286. Pennywitt, Kirk E. "Robotic Tactile Sensing." *Byte* 11 (January 1986), 177-178, 180, 182, 184-186, 188, 190, 193, 195-196, 198-200.
Discusses tactile sensing, and provides a list comparing various manufactured robot tactile sensors and their characteristics. Illustrated with drawings.

287. van Pelt, Dina. "Robotic Research Is Getting Touchy." *Insight* 4 (August 8, 1988), 52.
Reports on tactile sensors in robot fingers being developed at the Georgia Technical Research Institute, Georgia Institute of Technology, Atlanta, Georgia.

Vision

288. "A Neural Robot Learns to See." *Science News* 134 (July 9, 1988), 30.
Reports on a robotic system designed to teach a robot to see, developed by Michael Kuperstein, of Neurogen Laboratories in Brookline, Massachusetts.

289. "Improving Robot Vision." *USA Today* 115 (June 1987), 10.
Reports on robotic vision research conducted by Gerard Medioni of the University of Southern California, Los Angeles.

290. "Laser Eyes for Industrial Robots." *New Scientist* 96 (October 28, 1982), 234.

Describes a laser range-finder which robots could use, developed by the French government's INRIA. Illustrated with photograph.

291. "Robots That See with Sound." *The Futurist* 18 (October 1984), 61-62.

Describes Sonovision, an ultrasonic, wave-based vision system for robots, invented by Shawn Buckley, available through Coakley Heagerty Companies, Limited, Santa Clara, California. Illustrated with photograph.

292. "Seeing Robots Hold Key to Factory of the Future." *The Futurist* 15 (August 1981), 10-11.

Discusses robot vision systems, how they work, and how they might be used. Illustrated with photographs.

293. Dunbar, Phil. "Machine Vision." *Byte* 11 (January 1986), 161-162, 164, 166, 168, 170-173.

Discusses problems in the development of vision-system hardware, and describes some of the technology that has emerged to address problems. Includes a video glossary and list of commercially available systems and their features. Illustrated with drawings.

294. Fuerst, Irene. "R2D2 Meets G'Wrench." *Datamation* 31 (May 1, 1985), 50, 52, 55.

Reports on growth of the use of robotic vision systems in manufacturing.

295. Hapgood, Fred. "Inside a Robotics Lab: Looking for Stereo Vision." *Technology Illustrated* 3 (June 1983), 44-46, 48-49.

Discusses research on depth perception at the Massachusetts Institute of Technology, Cambridge. Illustrated with photographs and drawings.

296. Hollingum, Jack. *Machine Vision*. Kempston, Bedford, England & Berlin: IFS (Publications) Ltd. & Springer-Verlag, 1984. 113 p.

Covers what machine vision is, why to use it, how to get started, finding applications, working with a supplier, implementing a system, the future, and case studies. Illustrated with photographs and drawings.

297. Horn, Berthold K. P., and Ikeuchi, Katsushi. "The Mechanical Manipulation of Randomly Oriented Parts." *Scientific American* 251 (August 1984), 100-106; 108; 110-111.
Discusses various aspects of robot (machine) vision systems, and describes a machine vision system that can determine the position of a part with an arbitrary shape in a randomly arranged pile. Includes drawings and photographs.

298. Lerner, Eric J. "Robotics; The Birth of Vision." *Science Digest* 93 (July 1985), 28-35, 76-77.
Reports on work being done and progress in the development of computer vision systems for robots, especially at Carnegie Mellon University, Pittsburgh, Pennsylvania. Illustrated with color photographs.

299. Lowe, David. "Vision Leads Robots from the Factory." *New Scientist* 115 (September 10, 1987), 50-51,54.
Discusses vision in humans and how computer vision systems for machine vision work. Illustrated with photographs and drawings.

300. McKean, Kevin. "Computers That See." *Discover* 5 (September 1984), 72-73.
Reports on Neptune, a mobile robot created at the Robotics Institute of Carnegie Mellon University, Pittsburgh, Pennsylvania, that moves under guidance of its own vision, provided by two videocameras. Illustrated with photographs.

301. Myers, Edith. "Machines That See." *Datamation* 29 (November 1983), 90, 95, 98, 103.
Reports on machines and robots that use vision systems. Illustrated with photographs.

302. Port, Otis. "'Sunglasses' to Help Robots See Only What They Need to See." *Business Week* no. 2967 (October 6, 1986), 95.
Reports on shades developed by G. Michael Morris at the University of Rochester, New York, to screen out extraneous optical images "seen" by industrial robots using vision systems. Includes color drawing.

303. Rheingold, Howard. "The Well-Tempered Robot; Can It Learn to Behave?" *Psychology Today* 17 (December 1983), 38, 40-41, 44.
Discusses visual and mobility systems for robots. Considers the problems involved in developing robots with simple visual navigation. Includes photographs.

304. Rothfelder, Jeffrey. "Is There Intelligent Life in the PC?" *PC Magazine* 5 (January 14, 1986), 139-148.
Sidebar in article examines experiments with computer vision and robotics.

305. Smay, V. E. "Seeing-Eye Robot Makes Better, Faster Welds." *Popular Science* 223 (October 1983), 16.
Reports on welding robot, vision-and-control system unveiled at General Electric in Schenectady, New York, that can produce welds that match or surpass those of expert human welders. Includes photograph.

306. Watts, Susan. "Cross-Eyed Robots Start to See Like People." *New Scientist* 119 (September 15, 1988), 40.
Reports on robot vision research being conducted at the University of Rochester, New York. Illustrated with photographs.

307. Yeaple, Frank. "Robotic Tuner Has Remote Eye for Adjusting Hidden Pots." *Design News* 44 (February 1, 1988), 70.
Describes a robot that peers into holes and accurately locates the slot in an adjusting screw in the hole, developed by Anorad Corporation, Hauppauge, New York. Illustrated with color photographs and drawing.

308. Young, Dennis A. "Novel Silicon-Based Materials Could Lead to Robotic Vision." *Research & Development* 28 (February 1986), 56.
Reports on research on robotic vision being conducted at Bradford University, Yorkshire, England.

Voice

309. "Camera Obeys 'His Master's Voice.'" *Design News* 43 (February 9, 1987), 49, 51.
Reports on robot camera that operates by voice command, developed by the Advanced Technology Laboratories, Radio Corporation of America (RCA), Morristown, New Jersey.

310. "Robots Give Students at University of N. M. Hands-on Experience." *Technological Horizons in Education Journal* 15 (May 1988), 48, 50.
Describes Mohsen Shahinpoor's Mechanical Engineering Department, at the University of New Mexico, Albuquerque, and talking robots which use computer-installed, speech-recognition boards as controllers and a voice system to reply to commands.

CAREER INFORMATION

311. "Robotalks: A Strange Encounter in Which High Tech, Job Futures, and a Changing World Are Debated." *Career World* (September 1984), 20-21.
Provides information on career opportunities in the field of robotics.

312. "Robotics." *Industrial Education* 73 (May/June 1984), 10.
Provides an overview of industrial use and implications for job seekers.

313. Bone, Jan. *Opportunities in Robotics Careers*. Lincolnwood, IL: VGM Career Horizons, National Textbook Company, 1987. 147 p.

Covers background information on robotics, how a robot works, automation in manufacturing, robots in industry, machine vision, simulation and artificial intelligence, nonindustrial (service) robots, personal qualities needed for robotics career, education and training, job-hunting tips, information for women and minorities, international opportunities, associations, and periodicals. Includes glossary. Illustrated with photographs.

314. Cardoza, Anne, and Vlk, Suzee J. *The Robotics Careers Handbook*. New York: Arco Publishing, Inc., 1985. 154 p.

Provides an introduction to robotics, and covers education and training required, types of jobs available, salaries, military robotics, computer-integrated manufacturing, where the jobs are geographically, job search strategies, teaching jobs in robotics and educational robotics, how to select the right school, and job prospects. Includes lists of robotics associations and robotics-related publications, and bibliography. Illustrated with photographs.

315. Gite, Lloyd. "Here Come the Robots." *Essence* 14 (March 1984), 27.

Discusses robotics jobs, training required, and salary range, and includes interview with Leisha McKay, a robotics engineer. Illustrated with photograph.

316. Karmin, Monroe W. "Jobs of the Future." *U. S. News & World Report* 99 (December 23, 1985), 40-45.

Includes section on jobs in robotics field.

317. Lee, Mary Price, and Lee, Richard S. *Exploring Careers in Robotics*. New York: The Rosen Publishing Group, Inc., 1984. 128 p.

Covers what robots are, history of robots, robotics at the Massachusetts Institute of Technology, Cambridge, artificial intelligence, industrial robots, their applications and employment implications, robots in medicine and space, robotics as a career, academic preparation necessary, and the future of robotics. Includes outline of a robotics course. Illustrated with photographs.

318. Marrs, Texe W. *Careers with Robots*. New York: Facts on File Publications, 1988. 213 p.
Covers information on the robot revolution, charting a career in robotics, job choices, and training and education, and lists major robotics employers, higher education robotics programs, organizations, a glossary, and recommended reading. Illustrated with photographs.

319. Ross, Steven S. "Robots in the U. S. A." *Business Week's Guide to Careers* 4 (Spring/Summer 1986), 78-81.
Discusses uses of robots, and types of jobs available where robots are manufactured or used, and mentions some college and noncollege courses and training programs available. Includes color drawing.

320. Thro, Ellen. *Robotics Careers*. New York: Franklin Watts, 1987. 111 p.
Provides an introduction to robotics, and covers opportunities in the field in general, and in the following areas: designing new robots, hiring robots for a factory, planning the use of robots, programming robots, servicing robots, selling robots, inventing robots of the future, running your own business, and robotics in oceanography and space science. Illustrated with photographs.

SECTION 3. EDUCATIONAL ROBOTS

This section covers material on the integration of the study of robots into the school curriculum, robots designed primarily for use in an educational setting, classroom uses of educational robots, classroom activities in which a robot is a feature, and classroom materials for use when studying about robots.

Robots and robotics should be included in today's school curriculum just as computer science is. Not only is it important for young people to learn about robotics as a possible career option, but it is also important that they understand what a robot can and cannot do and the present and possible future impact of robots on our society. Many schools have included robotics in their curriculums and are offering robotics courses. The materials listed will serve as a guide to more information about such programs.

Many robots and robot kits have been specially designed for use in the classroom. Articles and books in the Educational Robots grouping provide information on some of the robots that are or have been available. Robots can be bought in hobby stores, toy stores, specialty or electronics stores, and through mail order catalogs dealing in educational robots, educational materials, electronics, computer software, audiovisual materials, or unusual objects, gifts, and gadgets. Some robots can only be purchased through the manufacturer or distributor. Robots no longer being manufactured can sometimes be found in stores dealing with remainder stock, at yard sales, and in second-hand stores. A listing of manufacturers and distributors of educationally useful robots is given in Appendix C, and a listing of educationally useful robots appears in Appendix D.

Robots are exceptionally motivating to young people. They are fascinated by the real robot as well as by a reasonable facsimile of a robot. A robot in the classroom can demonstrate phenomena that are not easily shown, or illustrate a lesson in a new way, while keeping the students' rapt attention. The materials listed in the Classroom Uses groupings provide a wide overview of the many ways teachers are using robots in their classrooms.

The groupings on classroom activities contain references to actual teaching activities that can be followed by any teacher. Ideas can be gathered by reading the annotations for all levels since many of the activities can be modified for a grade level other than the one using the activity described.

The Classroom Materials groupings contain both nonfiction and fiction types. Fiction was arbitrarily placed in a level that seemed most suitable. Many secondary-grade-level stories will also be enjoyed by readers in elementary grades. Some elementary-grade-level stories will be enjoyed by older readers. Annotations may help the user decide on a book's appropriateness for a specific class if the book itself is not available for review.

An additional source of classroom material on robots is the picture of a robot featured on a magazine or periodical cover. To help the teacher easily find these covers, they are described in annotations throughout this sourcebook, and a listing of all covers can be found in the index under "cover."

CURRICULUMS AND CURRICULUM CONSIDERATIONS

321. Blaesi, LaVon, and Maness, Marion. "Robots: An Impact on Education." *Technological Horizons in Education Journal* 11 (March 1984), 100-104.

Discusses robot types, concerns of the work force, incorporating robotics in educational systems, public school and postsecondary involvement, industry-education partnerships, and funding sources. Includes photographs.

322. Decker, Robert, and Krajewski, Robert J. "For Robotics, Slow-Scan Video, Satellite Beams, and More...the Future Is Now." *The American School Board Journal* 173 (March 1986), 32-34, 44.

Includes discussion on the introduction of robotics into the curriculum, with examples of schools that have robotics in their curriculum, and things to keep in mind before beginning a robotics program in a school.

323. DeVault, M. Vere; Rouiller, Kenneth; and Siewert, Dean. "Robotics and the Curriculum: Addressing the Questions." *Curriculum Review* 25 (March/April 1986), 32-34.
Describes the Schoolhouse Robotics program designed by educators at the University of Wisconsin - Madison to help students understand computer circuitry, and discusses curriculum implications of robotics and robot use. Illustrated with schematic.

324. DeVore, Paul W. "Microprocessors, Robotics, and Work." *Man/Society/Technology* 41 (January 1982), 6-9.
Explores technological developments, including robotics, which will have an impact on future technical education.

325. Dillon, Linda Schnulle. "Some Thoughts on Robots." *The Technology Teacher* 44 (January 1984), 13-14.
Covers history, what a robot is, applications, advantages of using robots, the use of robots in Japan and Soviet Union, changes robots may bring, and implications for industrial arts.

326. Fluty, Robert. "Robotics' Role in High-Tech Education." *Robotics Age* 6 (September 1984), 32-34.
Reports on interview with John Lawson, president of Feedback, Incorporated, Berkeley Heights, New Jersey. Discusses need for trained robotics personnel, the robotics curriculum, and criteria to consider when selecting an educational robot for industrial training, including credibility, flexibility, technology, safety, extendability, and support. Illustrated with photograph and drawing.

327. Fraser, Jeannette L.; Unger, Paul V.; and Lewis, Morgan V. *Robotics and Office Automation: Implications for Vocational Education*. Columbus: Ohio State University, National Center

for Research in Vocational Education, 1984. 30 p. ERIC document ED 240 269.

Reports on potential demand for skills in robotics. Includes information on training implications for vocational education, and skills and training requirements.

328. Gray, Robert A. "Robots: A High-Tech Tool for the Learner." *TechTrends* 33 (October 1988), 45-48.

Discusses robotic literacy, defined as knowledge of the advantages and limitations of the machine, basic operation, and useful applications. Describes industrial robots, provides an overview on how robots are being used in education - as a content area, an object of study, and as prosthetic aids in the special education field. Describes robot hardware systems, provides a model teaching strategy, and lists guidelines to facilitate effective robot use in the classroom.

329. Hamm, Jerry L., et al. "1984's School Enrollee." *Catalyst for Change* 13 (Winter 1984), 12-13.

Describes a four-phase program integrating robotics into the school curriculum in Coffeyville, Kansas: (1) Awareness; (2) Simple Programming; (3) Advanced Programming; (4) Application of Robotics in Society. The robot Hero 1 was used as the program's robot for study.

330. Hunter, Beverly, and Aiken, Robert. "The Evolution of Computer Courses in Vocational Education: Perspectives and Directions." *Technological Horizons in Education Journal* 14 (March 1987), 54-57.

Discusses the changing job market, robotics job training, the extent to which training programs for computer and other automation jobs are being implemented, especially in high schools, and future directions.

331. Long, Sandra M., and Long, Winston H. "Robotics: The Fourth 'R'?" *World Future Society Bulletin* 18 (March/April 1984), 15-19.

Discusses educational implications of the use of robots in home and factory.

332. McConnell, Barry. "Robots in the Curriculum." *Electronic Education* 4 (October 1984), 9-10.

Discusses place of robots in the educational curriculum, at the vocational-technical school, community college level. Cover of this issue shows robot toy in color.

333. Meys, John W. "Add Robotics - Without Much Expense." *The Technology Teacher* 47 (February 1988), 25-28.

Covers what students will learn from incorporating a robot into the curriculum, terminology, and construction of syringe-powered robots. Illustrated with photographs.

334. Seaman, Virgil A., and Steck, Francis X. "Robotics Applications for the Curriculum to Reflect Technology." Paper presented at the annual meeting of the American Industrial Arts Association, San Diego, California, March 30, 1985. 33 p. ERIC document ED 255 654.

Contains suggestions for integrating robotics into technology education courses at the elementary through high school levels.

335. Shane, Harold. *Teaching and Learning in a Microelectronic Age*. Bloomington, IN: Phi Delta Kappan, 1987. 86 p.

Includes "The Robot in Industry and Education: Establishing New Relationships," which covers the robot in myth and history, what a robot is, some implications for education, and concluding comments.

336. Sharon, Dan. "Robotics and Automation Literacy Project." *Industrial Education* 76 (February 1987), 44-46.

Discusses robot needs, goals of technology education, an educational model, a robotics literacy course, and why a robotics course should be taught. Illustrated.

337. Shur, Paul, and Lobello, Sharon. "Robots: The Next Educational Challenge." *Electronic Education* 3 (November-December 1983), 10-11, 15-16.

Discusses uses of robots, and potential for change and educational challenge. Illustrated with photographs.

338. Warnat, Winifred I. "Bridging the Gap in Robotics Education." Paper presented at the International Symposium on Industrial Robots/Robots 7, Chicago, Illinois, April 17, 1983. 15 p. ERIC document ED 268 973.

Discusses where robotics education is now provided, relevancy of current schooling for robotics training, and future skill needs.

339. Warnat, Winifred I. "Robotics: A Bridge for Education and Technology." Distinguished Lecturer Paper for the Michigan Department of Education, Lansing, Michigan, March 3, 1983. 35 p. ERIC document 227 087.

Provides historical overview of the robotics industry, and the role of robotics in education in relation to employment opportunities.

ROBOTICS COURSES AND SPECIFIC CURRICULUMS

General

340. *Robotics Curriculum*. Springfield, OH: Hearlihy & Company. 16 p.

For grades 5 through 12. Designed as companion to WAO and Avoider robot kits. Includes 30 student booklets, 150 two-page tests, teacher's lesson plan, answer sheet, press release. Covers history of robotics, terminology, robot capabilities, and future industrial and commercial uses.

341. Solomon, Gwen. "Learning to Use the Tools of the Future." *Electronic Learning* 9 (February 1990), 14-15.

Describes the Future Tech Lab in Southern Lehigh School District, Center Valley, Pennsylvania, where students at every grade level learn about robotics, fiber optics, holographic research, computer-aided design, space studies, and artificial intelligence. Illustrated with drawings and photograph.

Elementary Grades Level

342. Friedman, Madeleine. "Attention All Humans: Robotics Literacy Captivates Elementary Students." *TechTrends* 31 (January 1986), 12-14.

Describes a six-week, robotics literacy course for children ages 8 to 12, conducted by Broward Community College, Pembroke Pines, Florida. Lists robots and kits used in program, with teaching applications and comments, including Armatron, Big Trak, Capsela 1000 and Capsela 2000, fischertechnik Start 100 Kit and Motor and Gear Kit, Hero 1, Movit Kits (Circular, Medusa, Mr. Bootsman), Off-Roader Programmable (Radio Shack), Tomy series (Omnibot, Verbot, Flipbot, Dingbot), Robotix R-1000 and Robotix R-2000, and Turtle Tot. Cover of this issue shows a homemade robot in color.

343. Greene, Karen, and Smith, Dee. "Robots, Gobots, and a Peek at the Future." *Early Years* 17 (October 1986), 62, 64-66.

Describes agenda, activities, and projects of nine-week unit on robotics for third and fourth grades at Convenant Christian School, Greensboro, North Carolina, taught as part of the computer curriculum. Illustrated with photographs and drawing.

344. Jones, Preston K. "A Mini-Curriculum for Robotics Education." Ed. D. Practicum, Nova University, 1987. 45 p. ERIC document ED 290 448.

Presents a robotics curriculum for fourth and seventh grade students. Provides general objectives, specific lesson objectives, lesson plans, and evaluation results.

Secondary Grades Level

345. "With Cooperation from the Auto Industry, Robots Invade Michigan High Schools." *Phi Delta Kappan* 64 (October 1982), 147.

Describes a 64-hour summer course on industrial robotics conducted for juniors and seniors at the Northeast Oakland Vocational Education Center, Pontiac, Michigan, believed to be the first course of its kind in United States high schools.

346. Ascension Parish School Board. *Robotics in Industrial Arts. Final Narrative Report for the Exemplary Project.* Donaldsonville, LA: Ascension Parish School Board, 1984. 44 p. ERIC document ED 262 182.

Describes robot program in electronics classes at St. Amant High School, in Ascension Parish, Louisiana.

347. Auer, Herbert J. *Robotics: Instructional Manual. The North Dakota High Technology Mobile Laboratory Project.* Grand Forks: University of North Dakota, Department of Industrial Technology, 1985. 246 p.

Instructional manual containing 20 learning activity packets for use in a workshop on robotics. Includes illustrations.

348. Baker, Nick. "Computer Feed-Out." *Times Educational Supplement* no. 3616 (October 18, 1985), 40.

Reports on computer control and robotics courses at Lea Manor High School in Luton, England. Includes photographs.

349. Barker, Ronald G. "Types, Utilization and Application of Robots for Industrial Arts Education." *Industrial Education* 73 (May/June 1984), 12-14, 17-18.

Presents an introduction to robotics, including definition, how robots function, and applications in industry, and applications for the industrial arts laboratory. Illustrated with photographs and drawing.

350. Benton, Oliver, and Branch, Charles W. "Robots, Jobs, and Education. State-of-the-Art Paper." Knoxville: University of Tennessee, Office for Research in High Technology Education, 1984. 55 p. ERIC document ED 254 621.

Describes industrial robot characteristics, and practicalities of robot applications, considers changes that might happen to robotics over the next few years, and makes recommendations for robotics education at secondary vocational education and two-year college levels.

351. Besancon, Francis E. *A Secondary-Level Curriculum in Industrial Electronics and Robotics. Final Report.* Terryville, CT: Plymouth Board of Education, 1984. 192 p.

Describes development of a curriculum to provide, to students at secondary and adult vocational level, skills necessary to operate and repair industrial robots. Includes the robotics curriculum.

352. Cheng, Alex. "What Automation Training Does Today's Work Force Need?" *Industrial Education* 79 (January 1990), 12-15.

Describes Programmable Automation Literacy Lab at Motorola, robotics program at Pinckney High School, and Electronic Technology program at Mesa College, San Diego, California, all of which include robotics.

353. Cohen, Karen C., and Meyer, Carol D. "Robotics: Using Technology to Teach New Technologies." *Technological Horizons in Education Journal* 12 (August 1984), 88-92.

Describes a 32-hour curriculum of 21 units of instruction on industrial robotics, developed by Continuous Learning Corporation, Cambridge, Massachusetts.

354. Edmison, Glenn A., et al. "Robotics and Industrial Arts." Paper presented at the American Vocational Association Convention, Anaheim, California, December 1983. ERIC document ED 237 781.

Maintains that robotics should be taught in high school industrial arts programs as a major curriculum component.

Explains benefits of such a course. Includes ways robotics can be incorporated into junior high school industrial arts courses.

355. Gray, Lovett S. "And Now, a Major in Robotics." *New York Times* 135 (April 13, 1986), Ed 12.
Reports on robot programs in high schools. Illustrated with drawing.

356. Gromacki, Gerald, and Hannemann, Jim. "Getting Started in Robotics: An Update on the Nation's First Secondary-Level Robotics Program." *School Shop: Tech Directions* 49 (August 1989), 13-15.
Describes robotics program at Northeast Oakland Vocational Center, Pontiac, Michigan, and provides a list of equipment and costs. Illustrated with photographs.

357. Heath Company. *Industrial Robotics and Automation*. Benton Harbor, MI: Heath Company.
For high school and up. Course designed as companion to Robot Arm Trainer (Heath Company). Includes workbook and instructor's guide. Can also be ordered as home study course with optional final exam to earn four Continuing Education Units (CEU's). Covers components of industrial robot, role of microprocessor in industrial robotics, physical and operating characteristics of robots and servo control systems, components of hydraulic and pneumatic power supplies, programming a robot using on-line and off-line methods, and experiments with Robot Arm Trainer.

358. Heckard, Margaret. "Career-Minded High School Students Are Commanding Robots." *TechTrends* 31 (January 1986), 15-17.
Describes the robotics program of Arlington Public Schools, Arlington, Virginia, involving specialized courses in the system's Career Center. Includes photographs. Cover of this issue shows a homemade robot in color.

359. Hinson, Ray. "Training Programs Are Essential for Robotics Success." *Industrial Engineer* 15 (September 1983), 26, 28, 30.

Discusses training programs, categorizing training under the topics of awareness, justification, application, operation maintenance, and safety. Illustrated with photographs and drawing.

360. Hsiung, Steve C. "Designing a Robotics Curriculum." *Industrial Education* 26 (November 1986), 37-38.
Discusses design of an industrial robotics curriculum.

361. Kaiser, Joe. "Robots and Industrial Education." *Industrial Education* 72 (February 1983), 14-16.
Considers what a robot is, robot tasks, what industrial education students should learn about robots, and the future of robots. Illustrated with photograph.

362. Kansas Vocational Curriculum Dissemination Center. *Robot Service and Repair. Teacher's Guide*. Pittsburg, KS: Pittsburg State University, 1984. 280 p. ERIC document ED 250 517
Teacher's guide for course on robot service and repair. Covers introduction to robots, robot control systems, and service and repair. Includes lesson plans. Illustrated with drawings.

363. Kansas Vocational Curriculum Dissemination Center. *Robot Service and Repair. Student Guide*. Pittsburg, KS: Pittsburg State University, 1984. 354 p. ERIC document 250 518.
Student guide for course on robot service and repair. Covers introduction to robots, power supply, robot control systems, and service and repair. Illustrated with drawings.

364. Lavoie, Jane. "Reading, Writing, and Now...Robotics." *School Product News* 22 (November 1983), 41-42.
Reports on robotics programs in high schools, colleges, and universities. Illustrated with photograph and drawing.

365. Miller, Rex. "Teaching Robots and Robotics." *Industrial Education* 77 (May 1988), 16-18.
Discusses reasons technology courses which emphasize robotics should be included in the school curriculum at the

junior high (middle school) level, provides suggested course content for an introductory course, outlines course features, and considers the "factory of the future," the work environment, and retraining. Illustrated with photographs.

366. Mullen, Frank. *Basic Robotics*. 1984. 12 p. ERIC document ED 253 708.
Provides a curriculum outline of instructional materials and information concerning resources for use in teaching a course on industrial robotics. Illustrated with drawings.

367. Peckham, Susanne. "Secondary Students Tackle Robotics." *School Shop* 42 (November 1982), 17-19.
Describes a 64-hour course on industrial robotics, conducted at the Northeast Oakland Vocational Education Center, in Pontiac, Michigan, for 11th and 12th graders. Includes photographs.

368. Pierno, Michelle. "Automation/Robotics in High School." *Industrial Education* 74 (March 1985), 20.
Describes robotics program at Seaholm High School, Birmingham, Michigan. Illustrated with photograph.

369. Stoner, Candy L. "Teaching Hi-Tech: Robotics." *School Product News* 23 (November 1984), 25, 28-29, 32-33.
Describes programs and uses of robots in high schools, vocational schools, and two-year colleges, and training programs for teachers and technicians. Illustrated with photographs.

370. Thompson, Robert. "Planning a Course in Robotics." *Industrial Education* 72 (February 1983), 15-16.
Describes types of industrial robots, their use, and implications for technical education. Illustrated with photograph.

371. Tremer, Walt. "One Small (Robotic) Step." *Educational Leadership* 41 (September 1983), 35-36.

Presents information on Southern Lehigh School District, Pennsylvania, robotics program for students from a county vocational-technical school and the high school gifted program. Illustrated with photographs.

372. Valenti, Anthony. "Teaching Robotics Technology to Secondary-Level Students." *School Shop* 46 (February 1987), 18-21.

Describes robotics program at Bronx High School of Science, New York. Illustrated with photographs.

373. Whittaker, Alex. "Robots Bring Commerce to the Classroom." *New Scientist* 107 (August 29, 1985), 29-32.

Describes robotics course offered at St. Richard Gwyn RC High School, in Flint, Clwyd, England. Gives examples of one morning's activities, the elements of the course, and help received from companies involved in advanced technology. Provides information on Trekker, a robot vehicle developed by students in the course and sold by the school. Includes photographs and diagrams.

374. Wojcikiewicz, Karl, and Johnson, Ron. *Robot Applications*. Benton Harbor, MI: Heathkit/Zenith Education Systems, 1984. Variable paging.

A course covering management considerations, environmental feedback, vision systems, tactile sensing, computer-aided manufacturing, and robot applications. Illustrated with drawings.

Two-Year Institutions and Higher Education (Relevant to High School Programs)

375. "Robotics Moves in at S. C. T. I." *Electronic Education* 4 (October 1984), 13-14.

Describes robotics program at Somerset County Technical Institute, a postsecondary institution in Bridgewater, New Jersey. Includes photographs.

376. Arnold, W. F., and Carpenter, C. J. *Robotic Arms. A Contribution to the Curriculum. An Occasional Paper.* London: Further Education Unit, 1984. ERIC document ED 246 240.

Examines ways of providing technician training in the operating principles of robotic arms. Includes suggested objectives, outlines of three course units/modules based on objectives, and a list of references.

377. Asfahl, C. Ray. "Curriculum Dilemmas for Robotics Educators." *Technological Horizons in Education Journal* 11 (April 1984), 98-103.

Discusses problems associated with robot courses and curricula at the postsecondary level. Includes photographs.

378. Ashley, William; Abram, Robert; Konopka, Edward; and Carrico, Lloyd R. *Preparing for High Technology: Robotics Programs*. Columbus: Ohio State University, National Center for Research in Vocational Education, 1983. 63 p. ERIC document ED 228 472.

Covers current trends and related training, and curriculum guidelines and specialties. Appendixes include lists of robot manufacturers and postsecondary programs, a bibliography of robotic technical papers, and "Robotic Training: A Comprehensive Plan," by Lloyd R. Carrico.

379. Bailey, Mikal. "Some Tips on Fitting Robotics into Your IE Curriculum." *School Shop* 45 (October 1985), 24-25.

Provides background information on robots. Discusses what an industrial arts student should understand. Includes information on how a robot works, what a robot does, and robotics trainers. Includes photograph.

380. DeBourcy, Harry C. "Technology Literacy Training - The ACTT Method." *Technological Horizons in Education Journal* 16 (May 1989), 66-69.

Describes program, including Computer Integrated Manufacturing: Robotics and Automation Literacy curriculum, at the Advanced Center for Technology Training,

Eastern Iowa Community College, Davenport, Iowa. Illustrated with photograph.

381. Doty, Charles R. "Robotics/Automated Systems Technicians." Paper presented at the meeting of the American Vocational Association, Atlanta, Georgia, December 6-10, 1985. 9 p. ERIC document ED 265 398.

Discusses resources to upgrade robotics programs in two-year institutions.

382. Further Education Unit, London, England. *Robotics. Guidance for Further Education. FEU/PICKUP Project Report*. London: Further Education Unit, 1985. 51 p. ERIC document ED 256 901.

Gives results of survey of personnel in companies involved with robots to determine training needed for robotics career, suggests guidelines for course design, reviews British robotics curriculums, discusses course support, and provides information on the robot, robot stations, applications, and curriculum objectives.

383. Hull, Daniel M., and Lovett, James E. *An Assessment of the State of the Art of Curriculum Materials and a Status Assessment of Training Programs for Robotics/Automated Systems Technicians. Task Analysis and Descriptions of Required Job Competencies of Robotics/Automated Systems Technicians*. Waco, TX: Center for Occupational Research and Development, Inc., 1984. 156 p. ERIC document ED 255 672.

Presents results of a research study on the state of the art of robotics/automated systems technician training offered in the United States.

384. Hull, Daniel M., and Lovett, James E. *Robotics/Automated Systems Task Analysis and Description of Required Job Competencies Report. Task Analysis and Description of Required Job Competencies of Robotics/Automated Systems Technicians*. Waco, TX: Center for Occupational Research

and Development, Inc., 1985. 223 p. ERIC document ED 255 673.

Discusses the task analysis of the robotics/automated systems technician, including job description, and comparison of required competencies with existing and new courses.

385. Hull, Daniel M., and Lovett, James E. *Task Analysis and Descriptions of Required Job Competencies of Robotics/ Automated Systems Technicians. Outlines for New Courses and Modules*. Waco, TX: Center for Occupational Research and Development, Inc., 1985. 74 p. ERIC document ED 255 671.

Describes six robotics and automated systems courses developed by the Robotics/Automated Systems Technician project.

386. Hull, Daniel M., and Lovett, James E. *Task Analysis and Job Descriptions for Robotics/Automated Systems Technicians. Final Report*. Vol. 1. Waco, TX: Center for Occupational Research and Development, 1985. 128 p. ERIC document ED 255 674.

Describes the Robotics/Automated Systems Technician (RAST) project, which developed a robotics technician model curriculum for use by state directors of vocational education and two-year college vocational/technical educators.

387. Hull, Daniel M., and Lovett, James E. *Task Analysis and Descriptions of Required Job Competencies for Robotics/Automated Systems Technicians. Final Report*. Vol. 2: *Curriculum Planning Guide*. Waco, TX: Center for Occupational Research and Development, Inc., 1985. 73 p. ERIC document ED 255 675.

Provides curriculum planning guide to help institutions evaluate their community's needs and capabilities to meet these needs, in the area of robotics/automated systems technician training.

388. Husband, T. M., ed. *Education and Training in Robotics*. London & Berlin: IFS (Publications) Ltd. & Springer-Verlag, 1986. 311 p.

Sections cover management considerations, training programs in industry, college and university programs, and national and international strategies. Provides lists of educational robot manufacturers and robotic education institutions, technician job competencies, and specialty course outlines. Includes drawings and photographs.

389. Johnson, Keith K. "Computers as Teachers: Automating Training." *Production Engineering* 32 (July 1985), 64-68.

Describes computer-based training program to teach workers what industrial robots can do and how they are used. Illustrated with photographs.

390. Johnson-O'Mara, Daniel. "Challenges in Education." *Industrial Education* 77 (May 1988), 29-30.

Describes Robotics/Automated Systems Technician program at Hawkeye Institute of Technology, Waterloo, Iowa. Illustrated with photograph.

391. Kimbler, D. L. "Engineering Robot Applications and Education." *Technological Horizons in Education Journal* 11 (May 1984), 122-126.

Discusses problems arising in education in meeting the demand for robotics personnel in industry, proposes potential solutions, and describes the robotics program at the Department of Industrial and Management Systems Engineering, University of South Florida, Tampa.

392. Luzerne County Community College. *Development of Articulated Competency-Based Curriculum in Automated Systems/Robotics Technology. Final Report*. Nanticoke, PA: Luzerne County Community College, 1987. 88 p. ERIC document ED 288 018.

Describes development of the first year of a competency-based curriculum in automated systems/robotics at Luzerne County Community College, Nanticoke, Pennsylvania.

393. Lynch, Edward J. "Robotics Technician Training at Macomb Community College." 1983. 15 p. ERIC document ED 228 385.

Describes the robotics technician training program at Macomb Community College, Warren, Michigan.

394. McQuay, Paul L. *Automated Manufacturing/Robotics Technology: Certificate and Associate Degree Programs*. Media, PA: Delaware County Community College, 1984. 28 p. ERIC document ED 244 675.

Describes the program to be offered at Delaware County Community College, September 1984. Provides descriptions of the courses in the program.

395. Miller, Kenneth W. "The Education of Robotics Technicians." *Industrial Education* 75 (November 1986), 35-36.

Describes robotics program at Staples Technical Institute, Staples, Minnesota.

396. Petiprin, Frank C. "What to Do When the Robot Arrives." *Mathematics and Computer Education* 20 (Spring 1986), 79-86.

Describes the introductory robotics course taught at Jamestown (New York) Community College, and provides information on robots used in the course. Includes photographs and drawing.

397. Rehg, James. "Automation Training Tools of the Future." *Technological Horizons in Education Journal* 13 (May 1986), 78-81.

Discusses impact of new automation on education in automated systems, flexible manufacturing cells, flexible manufacturing systems, and educational goals, considers use of educational robot workcells for factory automation training, and describes robot workcell used at Piedmont Technical College, Greenwood, South Carolina.

398. Rehg, James A. "Robotics Programs: Automation Training in Disguise." *Industrial Education* 74 (May/June 1985), 17-18, 35.

Discusses need for robotics training, general objectives of an automated manufacturing or robotics training program, and equipment required in a robotics laboratory. Material is taken from author's book, *Guidelines for Robotics Program Development*.

399. Robotics International of Society of Manufacturing Engineers, Education and Training Division. *Sample Curriculum Guide for Robotics Automated Systems Technicians*. Dearborn, MI: Robotics International of Society for Manufacturing Engineers, Education and Training Division, 1985. 75 p.

Covers guidelines for two-year, postsecondary programs in robotics/automated system technology, technician job description, curriculum model, and job competencies. Provides samples of two-year, postsecondary, and high school technical programs, and laboratory equipment recommendations. Includes list of robotics textbooks and periodicals, and lists of educational and industrial robot manufacturers and distributors.

400. Shahinpoor, Mohsen, and Singer, Neal. "A New Robotics Instructional Laboratory." *Technological Horizons in Education Journal* 13 (August 1985), 54-56.

Describes a robotics laboratory for undergraduate and graduate students at the University of New Mexico, Albuquerque. Includes listing of equipment and describes laboratory projects. Includes photographs.

401. Sharon, Dan, et al. "The ORT Open Tech Robotics and Automation Literacy Course." *Programmed Learning and Educational Technology* 24 (August 1987), 200-210.

Describes robotics and automation course developed by the Organization for Rehabilitation Through Training in England, to be offered through an open learning environment, and includes hardware and software requirements, educational model, design principles, and future developments.

402. Sharon, Dan; Harstein, Judah; and Fischer, Martin. "Robotics Literacy Curriculum." *Industrial Education* 76 (March 1987), 18, 20-21, 61.

Describes self-study, distance learning course in robotics literacy being developed by the World ORT Union, London, England.

403. Tang, Le D. "Competencies Identification for Robotics Training." Paper presented at the American Vocational Association Convention, Anaheim, California, December 5, 1983. ERIC document ED 240 279.

Reports on a study to identify competencies for training robot technicians. Competencies were categorized into basic skills and job functions. Based on responses to a questionnaire, maintenance was found to be considered the most important job function, followed by operation, installation and programming, and, finally, application.

404. Wesp, Richard, and Schnippert, William. "Robotic Education." *The Computer Instructor* 1 (February 1985), 26-31.

Reviews the field of robotics, past and present. Describes a nontechnical course on the capabilities and applications of robots taught by the authors at Elmira College, Elmira, New York. Includes photographs.

EDUCATIONAL ROBOTS

405. "A Leg Up for Logo Robotics." *Science News* 113 (June 4, 1988), 364.

Reports on a version of the computer programming language, Logo, developed by Stephen Ocko and Mitchel Resnick of the Massachusetts Institute of Technology Media Laboratory, Cambridge, to use with Lego building pieces to create controllable machines, including robots.

406. "Desktop Robot Provides Hands-On Robotics Training." *School Product News* 22 (May 1983), 56.

Reports on the RM-101 Movemaster robot arm from Mitsubishi Electric. Illustrated with photograph.

407. "Inside Big Trak." *Robotics Age* 2 (Spring 1980). Reprinted in *Robotics Age: In the Beginning*, pp. 172-173. Edited by Carl Helmers. Rochelle Park, NJ: Hayden Book Company, Inc., 1983. 241 p.
Describes Milton Bradley's Big Trak, a programmable toy vehicle. Includes photographs.

408. "Roaming Robots." *Penny Power* 7 (October/November 1986), 10-12.
Provides results of the testing of 11 robots by eight young people. Robots tested were Bumpbot, Chatbot, Crackbot, Dingbot, Dustbot, Flipbot, Jim the Politician, Spotbot, Steve the Butler, Verbot, and Voltron Lion Force Computer Control.

409. "Robotics on the Move in Small Parts Markets." *Purchasing* 93 (July 8, 1982), 15.
Describes Microbot TeachMover, designed for classrooms, and Motion Mate (Schrader Bellows). Includes photograph of TeachMover.

410. "The Robots Are Coming." *Media & Methods* 23 (January/February 1987), 6.
Provides information on Spectron Instrument Corporation awards for student projects or suggestions, and Spectron's robotics packages, and on courseware material on robotics available from Rhino Robots.

411. Anderson, Eric. "'Down Under' Robot Adds Dimension." *Electronic Education* 3 (November/December 1983), 48.
Presents impressions of Tasman Turtle and its use in educational settings.

412. Angiolillo, Paul, and Blueston, Mimi. "Now Even Lego Is Going High-Tech." *Business Week* no. 3012 (August 17, 1987), 40.

Reports on Lego products that link with Logo to make programmable toys, including toy robots. Illustrated with color photograph.

413. Brooks, Susan. "Robots on the Loose!" *Instructor* 96 (August 1986), 95.
Considers Dingbot, Omnibot, CompuRobot, Verbot, and Flipbot, and their uses in the classroom.

414. Brooks, Susan. "Toy Store Robots." *Classroom Computer Learning* 7 (May 1987), 46-48.
Discusses ways to use robots and robot-like toys found in toy stores, especially with Logo programs. Includes SmartBot, CompuRobot II, Robie Senior, and Verbot.

415. Buckland, Miram. "The 50-Minute Robot." *Science and Children* 23 (September 1985), 21-22.
Provides instructions for construction of a wooden robotic arm which can be built by sixth-grade students. Includes photographs and diagrams.

416. Bull, Glen, and Cochran, Paula. "Tail of a Turtle: The Valiant Effort." *The National Logo Exchange* 3 (May 1985), 2-3.
Describes Valiant Turtle. Includes photograph.

417. Clark, Nigel. "Robots for Learning and Fun." *Electronics & Wireless World* 91 (August 1985), 41-43.
Describes various types and brands of robots, including "toys," turtles, and robot arms. Includes extensive list of robot suppliers in England. Illustrated with photographs.

418. Davis, Dennis D. "Learner Friendly Robot Shows the Basics of Computer Control." *School Shop* 48 (August 1988), 14-16.
Describes robot designed and built by students at Northwest Iowa Technical College. Includes Commodore Pet microcomputer program. Illustrated with photographs, drawings, and schematic.

419. Ellickson-Brown, Jim, and Mimnaugh, Doris. "Valiant Turtle Robot." *The Computing Teacher* 13 (June 1986), 37-39.

Reviews Valiant Turtle. Covers description, distributor support, impact on the classroom, software and documentation, and learning to live with the Turtle. Includes publisher's reply and photographs.

420. Elmaghraby, Adel S., and Jagannathan, V. "Study of the Hero-1 Educational Robot." *CoED* 5 (July-September 1985), 202-204.

Analyzes Hero 1 and describes future research directions and Hero 1 limitations.

421. Ferrell, Keith. "Robotic Workshop." *Compute!* 10 (May 1988), 62, 64.

Describes Capsela control system called Robotic Workshop. Illustrated with photograph.

422. Futcher, D. "Teaching with Robots." *Educational Computing* (December 1984), 28-35.

Discusses why robots should be used in education, what a robot is, what robots can do, and how they operate, are programmed, and connected to a computer. Describes use of the BBC Buggy, Movits, the Computing Kit, Armdroid 1 Robotic Arm, Micro Pulse Buggy, K1 Buggy, Beasty Mobile Base, Beasty Arm, Teach Robot, Ogre 1 Robotic Arm, Cyber 310 Robot Arm, Atlas Robotics System, and Micro Grasp Robot. Illustrated with photographs and drawings.

423. Gollon, Barbara. "Educational Toys with an Electronic Edge." *Home Office Computing* 6 (December 1988), 94, 96, 98.

Includes information on LEGO TC Logo, and color photograph of the Apple Computer-interfaced learning system, from which robots can be built.

424. Harris, Joss J. "An A-Mazing Logo Experiment." *Hands On!* 6 (Summer 1983), 12-13.

Discusses a Logo turtle robot versus a Logo computer-screen turtle, and educational value of the robot, and includes sample programs.

425. Heath, Larry. *Fundamentals of Robotics: Rhino Lab Manual.* Terre Haute, IN: Indiana State University, 1984. 95 p.
Teacher's manual designed for use with Rhino robot with teach pendant. Includes curriculum outlines, lab planning, objectives, equipment, and instructor's notes for lab lessons.

426. Helmers, Carl. "Photo Essay: A First Glimpse at Gemini." *Robotics Age* 7 (February 1985), 12-13.
Describes the Gemini robot. Sequence of photographs shows the robot seeking its charging station, a maneuver the robot performs when its battery needs recharging.

427. Howe, Samuel. "It's 1984 and Robots Are in the Classroom." *Media & Methods* 20 (January 1984), 8-10, 14.
Describes and includes photographs of the following robots: Hero l, RB5X, Tasman Turtle, and the Rhino robot arm. Also mentioned are Turtle Tot and BOB.

428. Hsiung, Steve C. "Simple Program Teaches Point-to-Point Robotics." *School Shop* 47 (December 1987), 17-18.
Describes the method and computer program used with a Rhino XR robot to illustrate point-to-point robotics. Illustrated with photograph.

429. Hutwagner, Jackie. "Construction and Application of Robotics." *Industrial Education* 75 (September 1986), 16-17.
Describes robot designed and built by students at Berry College, Mount Berry, Georgia, using a process that could be used at the high school level. Illustrated with photograph and drawing.

430. Inigo, R. M., and Angula, J. M. "An Inexpensive Robotics Laboratory." *CoED* 5 (July-September 1985), 4-9.
Describes design and use of a computer-controlled robot manipulator.

431. Iverson, Wesley R. "Small Robots Find a Niche." *Electronics* 55 (October 20, 1982), 100-101.
Reports on Rhino Robots and Microbot, and uses made of their robots.

432. Klein, George. "Robotics Projects Made Simple." *Industrial Education* 75 (May 1986), 10-11.
Describes use of MAC 100 control system developed by Techno, Incorporated, with student-built robot, at Columbia University, New York. Illustrated with photograph.

433. Kovalchuck, John, and Hayes, Harold. "Interfacing Pneumatics to Robotics: An Inexpensive Alternative to Computers." *Industrial Education* 74 (March 1985), 18.
Describes use of robots, powered and controlled by air, to teach robotics at Western Illinois University, Macomb, Illinois, applicable to high school robotics programs.

434. Krauss, Frederick G., and Holzhei, Donald E. "Rehabilitating a Robot." *Industrial Education* 76 (December 1987), 23-25.
Describes repair of nonworking Unimate robot donated to Delta College, University Center, Michigan. Illustrated with photograph.

435. Lough, Tom, and Harris, Judi. "A Computerized LEGO Kit Has Students Building LOGO Motives." *Electronic Learning* 7 (March 1988), 38-39.
Reviews LEGO TC Logo, and tests of its use with preschoolers and high school physics teachers as well as others. Illustrated with drawing.

436. Markowitz, Mike. "Discovering Science." *A+* 5 (July 1987), 26-28; 30-32; 34-39.
Surveys products that add instrumentation to an Apple II computer to turn it into an interactive science laboratory. Comparison chart lists suitable courses, complexity, documentation quality, ease of setup, ease of use, overall rating, and comments. Among those included are Robotic

Workshop (Multibotics) and fischertechnik Training Robot (fischer America). Includes color photographs.

437. Marsh, George, and Spain, Tom. "Robots in the Classroom." *Electronic Learning* 3 (March 1984), 49-53, 112.
Answers the questions, "What is an educational robot?", "Why a robot?", "What's available in educational robots?", "How would a robot fit in your classroom?", and "What makes a robot run?" Includes photographs and specific information on F. R. E. D., Topo, XR-2, RB5X, Hero 1, and Big Trak.

438. McConnell, Barry. "Topo: A New Tool for Teachers." *Electronic Education* 3 (November/December 1983), 49-50.
Describes Topo and possible uses. Includes photograph.

439. Neddenriep, Terry. "Construct a Robot Simulator for Use in the General Shop." *School Shop* 47 (October 1987), 20-21.
Provides directions for building robot model to help teach robotics. Illustrated with photographs and drawing.

440. Onosko, Tim. "Enter the Movits." *Creative Computing* 10 (December 1984), 119-120, 123.
Describes the Movit kit series and some of the robots available, such as Memocon Crawler, Avoider, and Line Tracer II. Includes photographs.

441. Pliskin, Nava, and Redlich, Amos. "The Educational Benefits of Program Generators for Instructional Robots." *Educational Technology* 29 (June 1989), 38-42.
Describes the Robot Motion Generator, software that controls the TeachMover instructional robot, and its advantages over a teach box and ARMBASIC for teaching introductory robotics literacy courses.

442. Pountain, Nick. "Taking Control." *Byte* 12 (March 1987), 279-280, 283-284, 286.
Includes descriptive information on the RTX Robot Arm from Universal Machine Intelligence, Limited, for educational users.

443. Robillard, Mark J. "Heath's New Hero." *Radio-Electronics* 7 (May 1986), 83-84.
Examines Hero 2000, including mechanical features, robot brains, and programming. Includes photographs.

444. Robillard, Mark J. "Using a Purchased Robot in the Lab." *Radio-Electronics* 56 (October 1985), 88, 90.
Discusses how to use the Scorpion Mobile Robot, a kit manufactured and sold by Rhino Robots. Illustrated with photographs.

445. Rosen, Marian. "Lego Meets Logo." *Classroom Computer Learning* 8 (April 1988), 50-51, 54, 56-58.
Describes use of LEGO blocks and LEGO TC Logo with first through eighth grades in St. Louis, Missouri, including projects building computer-controlled LEGO robots. Illustrated with color photographs.

446. Rusk, Natalie. "Opening a Robot Laboratory in Your School." *Media & Methods* 26 (May/June 1990), 16.
Describes use of LEGO/Logo to build robots and gives ideas for their use.

447. Sheets, Everett. "An Inexpensive Robot for the Classroom." *The Technology Teacher* 43 (September/October 1983), 25.
Briefly describes the computer-controlled robot vehicle, Big Trak. Includes photographs.

448. Sheets, Everett. "Making an Easy, Inexpensive Robot in the Industrial Arts/Technology Lab." *The Technology Teacher* 44 (November 1984), 28-30.
Provides instructions for connecting Radio Shack Armatron robot to a Timex-Sinclair 1000 microcomputer. Illustrated with photographs and schematics.

449. Sheffer, Nola. "The Turtle as a Helpful Robot." *C.A.L.L. Apple* 7 (May 1984), 62-64.
Describes the robotic turtles, Tasman Turtle and Turtle Tot.

450. Shimabukuro, Gini. "A Class Act: Junior High Students, LEGO and Logo." *The Computing Teacher* 16 (February 1989), 37-39.

Describes uses of the LEGO TC Logo Kit, which uses the computer to drive projects built with LEGO bricks, including robots. Illustrated with drawings and photographs.

451. Sievers, Dennis. "A Robotics Round-Up." *The Science Teacher* 56 (November 1989), 56-58.

Discusses use of LEGO/Logo, Robotix building kit, Capsela robotics system, Movit kits, and the fischertechnik robotic training system in the classroom. Illustrated with photographs.

452. Slesnick, Twila. "Robots and Kids; Classroom Encounters." *Classroom Computer Learning* 4 (March 1984), 54-59.

Presents information on classroom use of the robots, Hero, Topo, and Turtle Tot. Gives source, computer compatibility, computer languages, cost, capabilities, and options available, for these robots as well as for RB5X, Rhino XR, and MiniMover robots. Illustrated with photographs showing robots and specific uses.

453. Smith, Iola. "Robotics Project." *Times Educational Supplement* no. 3636 (March 7, 1986), XVI.

Reports on pilot robotics project in primary schools in England, and use of Robotix building kit.

454. Spiteri, Charles J. "Electronics, Robotics and More." *Industrial Education* 75 (November 1986), 18-19.

Describes teaching robot Atlas and its support software and instructional materials, available from L. J. Electronics.

455. Stonier, Tom. "Valient Turtle - Emphasis on the Practical." *Electronic Education* 3 (November/December 1983): 51-52.

Describes Valient Turtle (Valiant Turtle) and possible uses. Includes photograph.

456. Thornburg, David D. *Every Kid's First Book of Robots and Computers*. Greensboro, NC: Compute! Books, 1982. 83 p.
Introduces, at an elementary level, ways for children to develop skills in computer programming and geometry through the use of the Big Trak robot vehicle. Instructions are appropriate for other types of programmable robot vehicles of this type which respond to incremental commands, such as the turtle robots. Illustrated with diagrams of programmed paths.

457. Tillwick, Mike. "Introduce Students to Robotics with an Inexpensive Robot Arm." *School Shop* 46 (May 1987), 16.
Describes use of Armatron to introduce robotics in eighth grade and gives step-by-step exercises. Illustrated with photographs.

458. Tyre, Terian. "Robotics: Training Systems in Step with Industry's Advances." *Technological Horizons in Education Journal* 16 (May 1989), 12, 16.
Describes robot training systems by Rhino Robots, Incorporated, TII Robotic Systems, Heathkit/Zenith Educational Systems, fischer america, incorporated, and D & M Computing.

459. Viau, Elizabeth Anne. "Toys, Robots and Touch Pads: Catalysts for the Imagination." *Media & Methods* 25 (September/October 1988), 113-114.
Discusses robots that are useful in the classroom and ways they can stimulate student observations.

460. Watt, Molly. "Ask Molly." *Teaching and Computers* 2 (September 1984), 14.
"Programming Robots" section comments on value of educational robots in teaching programming, inspiring discussions, and introducing the field of robotics, and comments on Turtle Tot, Tasman Turtle, Topo II, F. R. E. D., and Big Trak.

461. Williams, Gurney. "Heath's Hero." *Omni* 5 (January 1983), 32.

Describes Hero 1 robot. Illustrated with color photograph.

462. Yeaple, Frank. "Robotic Kit Is More Than a Supertoy." *Design News* 40 (September 17, 1984), 119.
Describes PZ 80 Micom, robotic arm kit by Hikawa, available from Compusa Corporation, that comes with computerized clown that plays xylophone, and programmable, miniature drill press table. Illustrated with photographs.

463. Zemke, Ron. "The Robots Are Coming! Training Tomorrow's High-Tech Workers." *Training* 20 (June 1983), 18-20.
Discusses what a robot is, industrial applications, nonindustrial types, and table top teaching robots. Describes training robots offered by Rhino Robots, Microbot, Amptronics, Energy Concepts, Feedback, and Heathkit/Zenith. Illustrated with photographs.

CLASSROOM USES

General

464. "Making a Robot Out of Rubbish." *National Geographic World* 141 (May 1987), 25-29.
Describes CHICO robot building project for Hispanic students, sponsored by University of Maryland, College Park, Maryland. Illustrated with color photographs.

465. Bitter, Gary G., and Gore, Kay. "Robots in the Classroom: Another of Tomorrow's Teaching Tools Today." *Computers in the School* 2 (Winter 1985-86), 15-20.
Describes uses of robots in a second grade classroom as well as in other areas of the school. Provides information on some of the educational robots available, with prices and suppliers' addresses.

466. Campbell, Sharon, and Sears, Tim. "More Robots in the Classroom." *The Computing Teacher* 16 (December/January 1988-89), 9-11.

Explains how Turtle Tot, a turtle robot, is used with students from first grade through junior high school to explore mathematics and scientific principles. Includes photographs.

467. Chertok, Michael R. "Building Toward Technological Literacy." *Technological Horizons in Education Journal* 17 (August 1989), 62-64.

Describes a robot-building workshop, using Medusa and Peppy Movit Kits, conducted for young people at The Computer Museum, Boston, Massachusetts. Illustrated with photographs.

468. Goldberg, Albert L. "Robotics: What's in It for Education?" *Educational Technology* 24 (August 1984), 34-35.

Considers the use of the classroom robot, the robot as symbol, and implications of use.

469. Howell, R. D.; Mayton, G. B.; and Baker, P. R. "Design Issues in the Use of Robots as Cognitive Enhancement Aids." *Educational Media International* 25 (June 1988), 104-105.

Describes research concerned with the educational use of robots by physically handicapped children, especially in providing early learning experiences in controlling and manipulating the educational environment.

470. Neighbours, David, and Slaton, Frank. "Robots in MY Classroom?" *The Computing Teacher* 14 (March 1987), 24-28, 45.

Covers teaching about robots, outlining four basic robotic principles for students (programming, mobility, power source, and mobility), and teaching with robots. Boxes provide information on commercially available robots, robotics reference materials, a critical thinking process activity with a robot/computer, and robot tasks for different grades. Includes information on CompuRobot, CompuRobot II, fischertechnik kits, Maxx Steele, Memocon Crawler, Omnibot, Super Tasman

Turtle, Tasman Turtle Tot, Valiant Turtle, and Verbot. Cover of this issue shows robot hand writing on chalkboard.

471. Post, Paul E.; Howell, Richard D.; and Rakocy, Lew. "Robot Technology: Implications for Education." *Educational Technology* 28 (January 1988), 39-45.

Covers what a robot is, what makes up a robot system, types of robots, programming, criteria for selecting a robot for education, and educational uses of robots as exemplars, objects of instruction, and functional prosthetic aids. Illustrated with drawing.

472. Schreiber, R. R. "The ABCs of Educational Robots." *Robotics Today* (February 1984). Reprinted in *Education and Training in Robotics*, pp. 267-271. Edited by T. M. Husband. London & Berlin: IFS (Publications) Ltd., & Springer-Verlag, 1986. 315 p.

Considers advantages of using educational robots in training and for robot literacy, educational packages available, and levels of training.

473. Terry, Colin, and Thomas, Peter, eds. *Teaching and Learning with Robots*. Beckenham, Kent, England: Croom Helm, Ltd., 1988. 227 p.

Includes Richard Pawson, "What Is the Role of the Personal Robot?"; David Wharry, "Control Technology and Robotics in Primary Schools"; David Argles, "Yes - But How Young?"; Alan Giles, "Electronic Sentences"; Alex Whittaker, "Robotics - An Evolving Syllabus for GCSE"; Mike Bostock, "Software Approaches to Computer Control"; Peter Nicholls, "Life Is Self Marking"; John Martin, "Conceptual Components of Control"; Paul Chung, "Learning Control Applications Through Programming"; John Billingsley, "Robots in the Classroom - The Practical Approach"; Dave Doyle, "Electronics to Robotics in Craft Design and Technology"; John Bonington, "An Educational Technologist's Perspective on the Process"; Ian MacLean, "Robots in Industry and Post-School Education"; and Ian Warnock, "Programmable Controllers in Robotics and Control."

474. Tyre, Terian. "Live TV Broadcasts from Ocean Floor Bring New Depth to Science Education." *Technological Horizons in Education Journal* 17 (August 1989), 42, 44-46.

Describes the JASON Project, a live broadcast to schools, of underwater exploration conducted by the Center for Marine Exploration, Woods Hole Oceanographic Institute, Woods Hole, Massachusetts, using remotely operated vehicles, *Jason* and *Medea*, in May 1989. Illustrated with photograph and drawing.

Preschool Level

475. Forman, George. "Observations of Young Children Solving Problems with Computers and Robots." *Journal of Research in Childhood Education* 1 (August 1986), 60-74.

Presents findings of a study of children, aged three to seven years, who were observed using computer graphics and robotics, with robots providing a problem-solving environment.

476. Keller, Joan, and Shanahan, Dolores. "Robots in Kindergarten." *The Computing Teacher* 10 (May 1983), 66-67.

Describes use of Big Trak to teach decisionmaking, estimation, divergent thinking, directionality, numerical concepts, and creative problem-solving skills.

Elementary Grades Level

477. "Marvel of the Bronx." *Time* 103 (April 1, 1974), 48.

Reports on Leachim, a robot built for fourth grade teacher Gail Freeman by her husband, Michael Freeman, and used to teach in the classroom at P. S. 106, Bronx, New York. Includes photograph.

478. "Push My Button." *Electronic Education* 5 (February 1986), 19, 28.

Reports on a pilot project involving use of a RB5X robot with students from 22 elementary schools in New Mexico, funded by Rio Grande Robotics, Incorporated, headed by Niki Delgado. Language arts and mathematics skills were among those taught using the robot. It was found that students retained a high level of the specific information presented in sessions with the robot, and a good deal of new-found self-esteem and enthusiasm was noticed among remedial students. Includes drawing.

479. Barnett, Harvey, and Better, Jennifer. "The Language of Robots." *Computers, Reading and Language Arts* 1 (Spring 1984), 35-37, 58.

Lists ways that incorporating a programmable robot into an elementary school curriculum improves students' speaking, listening, reading, and writing skills.

480. Delgado, Niki. "Robots in the Classroom." *Robotics Age* 6 (September 1984), 18-20.

Discusses reasons for using robots in the classroom. Stresses that robots are three-dimensional objects with which student can interact, students can instantly evaluate the results of their work when they program a robot, students gain a feeling of control over a tangible object in their environment, and students will readily use imagination to pretend that the robot's technological limitations do not exist. Describes a pilot classroom test with students ages seven through 13, using the RB5X robot.

481. Freeman, Michael, and Mulkowsky, Gary P. "Advanced Interactive Technology: Robots in the Home and Classroom." *The Futurist* 12 (December 1978), 356-357, 359-361.

Although dated, includes useful information on using robots in the classroom. Particularly illustrates uses of Leachim and 2-XL, educational robots invented by co-author Freeman. Includes photographs.

482. Grosswirth, Marvin. "Leachim the Teaching Robot." *Datamation* 20 (August 1974), 64-67.
Reports on the use of Leachim, a robot built for Gail Freeman by her husband Michael Freeman for use in her fourth grade class at P. S. 106, Bronx, New York. Explains how the robot is used, how it works, and what makes up its computer system. Includes photographs.

483. Lange, Mary. "Using the Turtle Tot Robot to Enhance Logo for the Hearing Impaired." *American Annals of the Deaf* 130 (November 1985), 377-382.
Describes uses of Turtle Tot and Logo in teaching. Illustrated with photographs and drawings.

484. Larson, Garrett W., and Haugerad, Al. "Robotics in Elementary Education." *Robotics Age* 6 (September 1984), 21-22.
Discusses uses of robots in the elementary classroom to teach basic robotics principles and computer literacy.

485. Shryock, Betty, et al. "Reading, Writing, and Robots." *Instructor* 94 (August 1984), 30-32, 34.
Outlines the history of robots, types of robots, and their uses and advantages, and suggests how to use this information in a unit on robots for grades one through six. Describes multidisciplinary activities to go with the unit. Describes RB5X, Topo, and Tasman Turtle, and provides information on the traveling robot exhibit of the American Craft Museum in New York City. Includes color photographs.

486. Stern, Sam, and Fichter, Eugene F. "Hands-on Robotics (For Just $8)." *The Technology Teacher* 45 (November 1985), 23-25.
Describes a robot building project used in a two-week summer class titled, "Robots and How They Work," for a group of talented and gifted fifth, sixth, and seventh graders. Robot used Big Trak motor assembly. A list of parts and supplies used, suggested supply sources and prices, is provided. Includes photographs and schematic diagram.

487. Watt, Molly. "The Robots Are Coming!" *Popular Computing* (Mid-October 1984), 71-72, 75-77.

Describes use of robots with children in various situations. Gives manufacturer, uses, standard features, documentation, options, and price for the following robots: F. R. E. D., Tasman Turtle, Turtle Tot, and Topo II.

488. Wood, Carol Hahn. "Project REM." *Science and Children* 20 (May 1983), 21-23.

Describes Project REM (Resources, Energy, Mankind) in six-year science curriculum which includes building Mr. REM, a "robot."

Secondary Grades Level

489. "Robots Displace Slide Rules From the Classroom." *New Scientist* 97 (March 10, 1983), 651.

Reports on an exhibit on school projects in electronic arts, including robots, in Birmingham, England, and on England's Microelectronics Education Programme.

490 "The Smartclassroom." *A+* 6 (October 1988), 29-31.

Describes an electronic learning center for seventh grade students at E. O. Green Junior High School in Hueneme, California. The first Smartclassroom is a paperless and textbookless educational environment. Robots used include two turtle robots, a fischertechnik Robotic Kit, and a Hearoid robot named Robie. Includes photographs.

491. Ehrlich, Raymond B. "Jack the Gripper and Robogames: A Robotic Mass Production Project." *The Technology Teacher* 47 (May/June 1988), 23-24.

Describes use of robotic arm to play computer games designed by students at a junior high school. Illustrated with photographs.

492. Glasthal, Jackie. "Tying Knots Around a Robot." *Teaching and Computers* 6 (October 1988), 11.

Reports on a robotics session of the Police Athletic League Logo Computer Camp, Bronx, New York. Includes photograph showing the robot Armatron being used to tie a shoelace.

493. Moore, J. L. "Development of a Questionnaire to Measure Secondary School Pupils' Attitudes to Computers and Robots." *Educational Studies* 11 (1985), 33-40.

Summarizes study which involved development and testing of a self-report questionnaire, applicable to all pupils between 13 and 16 years old. Article is based on author's Master of Philosophy thesis, "The Development and Use of a Questionnaire Measuring Secondary School Pupils' Attitudes to Computers and Robots," 1984, University of Nottingham, England.

493a. Moore, J. L. "An Empirical Study of Pupils' Attitudes to Computers and Robots." *Journal of Computer Assisted Learning* 1 (1985), 87-98.

Summarizes study which involved development and testing of a self-report questionnaire, applicable to all pupils between 13 and 16 years old. Article is based on author's Master of Philosophy thesis, "The Development and Use of a Questionnaire Measuring Secondary School Pupils' Attitudes to Computers and Robots," 1984, University of Nottingham, England.

494. Walsh, Edwina. "There Once Was a Robot Named..." *Electronic Education* 5 (February 1986), 22.

Describes a Technology Awareness Week of activities to celebrate the opening of a new school computer center. Among the activities were a high-tech poetry contest with entries of robot limerick, a robot talk-alike contest, and a robot demonstration. Includes example of a poetry entry with accompanying robot drawing.

CLASSROOM ACTIVITIES

General

495. "Byte-Size Ideas: In the Robot Room." *Classroom Computer Learning* 8 (February 1988), 77.
Describes an activity for teaching Logo by using a "robot room" with a grid of squares taped or drawn on the floor, through which a student makes robot-like moves according to Logo and other commands. Includes drawing.

496. "Direct a Robot." *Learning* 12 (September 1983), 208.
Describes an exercise to promote clear and explicit writing by having students write directions to be used by a robot to perform some action.

497. Adams, Dennis. "Programmers and Robots." *Instructor* 93 (January 1984), 128.
Describes classroom activity to show how robots are controlled by computer programs.

498. Bollinger, Richard. "Tarnic, the Smoking Robot." *Science Activities* 22 (April/May 1985), 35-37.
Provides instructions for constructing easy-to-make, cigarette-smoking robot, to demonstrate effect of cigarette smoke on water, and therefore on the body. Illustrated with drawing.

499. Cassetto, James M. "Teaching Robotics without Robots: Simple Method Introduces Students to the Basics." *School Shop* 49 (August 1989), 16.
Presents a list of robotic principles to cover, a sample programming system, and sample programming sheet.

500. Evans, Larry. *How to Draw Robots and Spaceships*. San Francisco: Troubador Press, 1982. 64 p.

Gives instructions on how to draw contemporary and humanoid robots, robot accessories and backgrounds, and robots as machines and villains. Illustrated with drawings.

501. McDowell, Debra, and Bell, Edna. "Haute Couture for the Erector Set." *Vocational Education Journal* 64 (October 1989), 69-70.
Describes class activity of clothing and textile and merchandising students at Southwest Missouri State University, Springfield, to design and make a 13-piece robot wardrobe for Prevention Pete/Safety Sue, a robot developed by Robotronics, Incorporated, Spanish Fork, Utah, to use as an educational aid explaining the importance of safety to children and adults.

502. Newman, Debby. "Robbie Robot Has Quilt; Will Cuddle." *Crafts 'n Things* 12 (May 1987), 12-13, 18-19, 36-38.
Provides instructions for making robot quilt and robot pillow/toy. Illustrated with photographs and drawings.

503. Westcott, Sandra, and Mager, Robin, eds. *Robots: A Teacher's Guide to Resources and Activities*. Pittsburgh, PA: Buhl Science Center, 1985. 28 p.
Contains projects, activities, and discussion questions for use at the elementary and middle school level, covering what a robot is, robot language, various robot uses, robots and society, image of a robot, technology and robots, and fun with robots. Also includes glossary, reading for young people, resources for teachers, and a list of manufacturers and distributors.

Elementary Grades Level

504. Braun, Charlotte. "Robots." *School Arts* 73 (June 1974), 42.
Describes a 6th grade project to build robots as part of a sculpture unit in an art class.

505. Farris, Cynthia Cox. "Focus on Outer Space." *Arts Activities* 101 (June 1987), 46-47.
Describes projects for primary grades, including making robots. Illustrated with photographs.

506. Farris, Cynthia Cox. "Let's Make Robots." *Arts Activities* 100 (January 1987), 54-55.
Describes ways to make robots in third grade art class. Illustrated with photographs.

Secondary Grades Level

507. "Integrated Manufacturing Systems: The Role of Robotics." *The Technology Teacher* 44 (January 1985), 15-21.
Discusses industrial robots in integrated manufacturing systems, and provides information on constructing robot conceptor, with a mathematics-science interface, designed as a high school class activity. Illustrated with photographs, drawings, and schematics.

508. "Upper Grades Ideas." *Classroom Computer Learning* 4 (September 1983), 72-73.
Includes teaching activity idea - a writing activity speculating on what would happen if robots did not follow Isaac Asimov's "Three Laws of Robotics."

509. Anderson, Rodney, and Carter, Lee. "Robot Construction Using Inexpensive Materials." *Industrial Education* 74 (November 1985), 22-23.
Describes challenge project for entire class to build a robot, using traditional materials found in most school laboratories. Illustrated with photographs.

510. Handrisanto, Ngallmin; Herrero, Gustavo; and Gibson, Roger. "Evolving Capabilities in Robotics." *Industrial Education* 78 (October 1989), 14-15.

Describes an activity to program robotic arm to differentiate among colored balls. Illustrated with photograph and drawing.

511. Hiester, Sue Moore. "Robotropolis." *School Arts* (March 1986), 14-17.
Describes a seventh grade class project to make a robot village, using found objects and photograms.

512. Neidlinger, William L. "A Robot Project for Your CAM Program." *Industrial Education* 78 (January 1989), 14.
Describes high school project using project in *How to Make Computer-Controlled Robots for Commodore 64 and VIC 20*, by Tony Potter. See number 781. Illustrated with photographs.

513. Oaks, Merrill M., and Richards, Harold. "Robotics in the Technology Education Curriculum." *The Technology Teacher* 48 (November 1988), 23-26.
Discusses advantages of the study of robotics, and methods for teaching robotics through classroom and laboratory construction activities. Illustrated with photographs.

CLASSROOM MATERIALS

General

514. "G.A.R.D.N.E.R.T.R.O.B.O.T." *Teaching and Computers* 1 (April 1984), 31.
Mini-poster. A drawing of a robot that pulls weeds and waters flowers, with detailed explanations.

515. *Kids Working with Robots: RB5X*. New York: Trillium Press.
"20+ activities designed to enable children to control the RB5X." Grades 5 up. Box 921, Madison Square Station, New York, NY 10159.

516. "Strip Starring Suburban Robot Starting Soon." *Editor & Publisher* 117 (December 29, 1984), 28.
Reports on introduction of cartoonist Jim Meddick's cartoon strip "Robotman."

517. Astor, David. "Over 200 Newspapers at the Age of 23." *Editor & Publisher* 118 (March 23, 1985), 46-47.
Chronicles history of cartoon strip "Robotman" and its author, Jim Meddick.

518. Deal, Walter F., III, et al. "Resources in Technology: Robotics; An Emerging Technology." *The Technology Teacher* 49 (February 1990), 15-22.
Covers what a robot is, different types, industrial, service, education and training, and special purpose robots, and math/science/technology interface. Includes construction activity, student quiz, and possible student outcomes. Illustrated with photographs and drawing.

519. Razzi, Jim. *Tyrannobot: Punch-Out Robot Dinosaur.* New York: Scholastic Inc. 32 p. ISBN 0-590-40790-2.

Preschool Level

Nonfiction

520. Gathrid, Jonathan. *Alphabots.* San Diego, CA: Slawson Communications, Inc., 1985. 28 leaves..
Devotes each page to a different letter of the alphabet. Letters are drawn as robots with each one holding or doing something with an object whose name begins with that letter. For example, robot G is eating grapes, robot Q is carrying a quilt, and robot Y is playing with a yo-yo. Capital and noncapital letters are also shown on each page. Illustrations are in color.

521. Hoffman, Phyllis. *The Robot Book.* Illustrated by Carol Hudson. New York: Scholastic Inc., 1986. 10 p.

A Colorforms Concept book. Introduces basic shapes and how they may be used to build a robot. Includes vinyl stick-on shapes that may be reused.

Fiction

522. *Surprise Robot!* Illustrated by Yasuko Ito. New York: Simon & Schuster, Inc., 1987. 8 leaves.

A sturdy book, all in color, about what robots can do, with moveable parts that pull out to form a robot with the book as its body.

523. Krahn, Fernando. *Robot-Bot-Bot.* New York: E. P. Dutton, 1979. Unpaged.

An all-picture, no-words book telling the story of a family robot that began to misfunction due to tampering with its inside mechanism, and how it was repaired and became a useful member of the family again.

Elementary Grades Level

Nonfiction

524. "Bionic Bafflers." *Instructor* 94 (August 1984), 38.

Provides three reproducible puzzles involving robots: two word puzzles and one coded, connect-the-dots puzzle.

525. Ratner, Ellen. "Robots on the March." *Instructor* 94 (August 1984), 39.

Provides the music and words for a reproducible song about robots.

526. Roth, Ilene, and Roth, Leah. "Is There a Robot in the House?" *Teaching and Computers* 5 (April/May 1988), 45-49.

A play to demonstrate the growing roles that robots are playing in our society and in our lives. Includes descriptions of costumes and props. Illustrated with drawing.

527. Sylvester, Diane. *Inventions, Robots, Future*. Santa Barbara, CA: The Learning Works, Inc., 1984. 112 p.
Workbook. Robots section includes information on robots, activities, pretest, and posttest. Illustrated with drawings.

Fiction

528. *Star Wars: The Mystery of The Rebellious Robot*. Illustrated by Mark Corcoran. New York: Random House, 1979. 29 leaves.
An adventure story about Star Wars characters, including Artoo Detoo (R2D2), See Threepio (C3PO), and others, and a problem of misfunctioning robots. Illustrated with color drawings.

529. Ames, Mildred. *Is There Life on a Plastic Planet?* New York: E. P. Dutton, 1975. 134 p.
Tells the story of a young girl who has a look-alike, life-size doll that takes her place at school and at home.

530. Asimov, Janet, and Asimov, Isaac. *Norby and the Invaders*. New York: Walker and Company, 1985. 138 p.
Relates the adventures of a teenage Space Academy student and Norby, a second-hand robot, on a strange planet.

531. Asimov, Janet, and Asimov, Isaac. *Norby and the Lost Princess*. New York: Walker and Company, 1985. 129 p.
Relates the adventures of a teenage Space Academy student and Norby, a second-hand robot, on a distant planet, helping to rescue a lost princess.

532. Asimov, Janet, and Asimov, Isaac. *Norby and the Queen's Necklace*. New York: Walker and Company, 1986. 136 p.

Relates the adventures of a teenage Space Academy student and Norby, a second-hand robot, transported back in time to France in 1785.

533. Asimov, Janet, and Asimov, Isaac. *Norby and Yobo's Great Adventure*. New York: Walker and Company, 1989. 100 p.
Tells the adventures of a teenage Space Academy student and Norby, a second-hand robot, as they travel back in time to trace the history of an admiral's family heirlooms.

534. Asimov, Janet, and Asimov, Isaac. *Norby Down to Earth*. New York: Walker and Company, 1989. 107 p.
Relates the adventures of a teenage Space Academy student and Norby, a second-hand robot, when they return to Earth and New York City.

535. Asimov, Janet, and Asimov, Isaac. *Norby Finds a Villain*. New York: Walker and Company, 1987. 102 p.
Relates the adventures of a teenage Space Academy student and Norby, a second-hand robot, traveling backward and forward in time, and to another universe.

536. Asimov, Janet, and Asimov, Isaac. *Norby, Robot for Hire*. New York: Berkley Publishing Group, 1987. 203 p.
Includes *Norby and the Invaders* and *Norby and the Lost Princess*.

537. Asimov, Janet, and Asimov, Isaac. *Norby, the Mixed-Up Robot*. New York: Walker and Company, 1983. 93 p.
Relates the adventures of a teenage Space Academy student and Norby, a second-hand robot, in New York City of the future. Illustrated.

538. Asimov, Janet, and Asimov, Isaac. *Norby Through Time and Space*. New York: Ace Books, 1988. 202 p
Includes *Norby and the Queen's Necklace* and *Norby Finds a Villain*.

539. Asimov, Janet, and Asimov, Isaac. *Norby's Other Secret*. New York: Walker and Company, 1984. 138 p.

Relates the adventures of a teenage Space Academy student and Norby, a second-hand robot, on a dragon-inhabited planet.

540. Asimov, Janet, and Asimov, Isaac. *The Norby Chronicles*. New York: Ace Books, 1986. 185 p.

Includes *Norby, the Mixed-Up Robot* and *Norby's Other Secret*.

541. Baker, Betty. *Worthington Botts and the Steam Machine*. Illustrated by Sal Murdocca. New York: Macmillan Publishing Company, Inc., 1981. 54 p.

A Ready-to-Read book. Tells the story of a boy who liked to read, and what happened after he built a robot steam machine to help him find time to read more. Illustrated with color drawings.

542. Beatty, Jerome, Jr. *Maria Looney and the Remarkable Robot*. New York: Avon Books, 1979. 144 p.

Tells the adventures of a family robot who is robotnapped and then saved by the girl whose life the robot once saved. Illustrated with drawings.

543. Black, Christopher. *The Android Invasion*. Illustrated by Maelo Cintron. New York: Dell Publishing Company, Inc., 1984. 117 p.

A space adventure in which the reader makes decisions that determine how the story unfolds. The reader is aided in the story by a task/operational robot, Model 2, called 2-Tor. Illustrated with drawings.

544. Brink, Carol Ryrie. *Andy Buckram's Tin Men*. New York: The Viking Press, 1966. 192 p.

An adventure story about Andy, his friend, Sparrow, his young cousin, Dot, and the four homemade boy and girl robots who become "alive" during an electric storm. Illustrated with drawings.

545. Bunting, Eve. *The Robot Birthday*. New York: E. P. Dutton, 1980. 80 p.

Tells the story of a boy and his birthday robot, made by his mother, an electronics teacher. Illustrated with color drawings.

546. Bunting, Eve. *The Robot People*. Illustrated by Don Hendricks. Mankato, MN: Creative Education, Inc., 1978. 25 p. Also available on sound cassette. See number 1504.

An adventure story about two robots, Link and Magnus, who work in an underground nuclear operation, and what happens when the son of a scientist working there learns that Magnus has developed a dislike for all humans. Illustrated with drawings.

547. Cone, Molly. *Mishmash and the Robot*. Illustrated by Leonard Shortall. Boston, MA: Houghton Mifflin Company, 1981. 120 p.

Tells the story of the daily adventures of a dog named Mishmash, including his encounters with a supermarket's advertising robot named Robot. Illustrated with drawings.

548. Cook, Ann, and Mack, Herb. *Robot Comes to Stay*. Illustrated by Irene Trivas. New York: Dell Publishing Company, Inc., 1982. 31 leaves.

First in a series about a girl and her robot named Robot. This book tells how the girl found a robot in a junkyard and repaired it. Illustrated with color drawings.

549. Cook, Ann, and Mack, Herb. *Robot and the Flea Market*. Illustrated by Irene Trivas. New York: Dell Publishing Company, Inc., 1982. 30 leaves.

Continues the story of Robot. Tells how Robot helped the children do their spring-cleaning by holding a flea market sale, and what happened as a result. Illustrated with color drawings.

550. Cook, Ann, and Mack, Herb. *Robot Goes Collecting.* Illustrated by Irene Trivas. New York: Dell Publishing Company, Inc., 1982. 30 leaves.

Continues the story of Robot. Tells what Robot did with all the items that he collected while helping the girl who owns him make a collection of balls after they visited a collectors' fair. Illustrated with color drawings.

551. Cook, Ann, and Mack, Herb. *Robot in Danger.* Illustrated by Irene Trivas. New York: Dell Publishing Company, Inc., 1982. 30 leaves.

Continues the story of Robot. Tells how Robot and the girl who owns him went shopping and bought plants, and what happened when Robot watered the plants. Illustrated with color drawings.

552. Cook, Ann, and Mack, Herb. *Robot Saves the Day.* Illustrated by Irene Trivas. New York: Dell Publishing Company, Inc., 1982. 30 leaves.

Continues the story of Robot. Tells how he helped the children practice for Sports Day and how he helped to make that day a success. Illustrated with color drawings.

553. Cook, Ann, and Mack, Herb. *Robot Visits School.* Illustrated by Irene Trivas. New York: Dell Publishing Company, Inc., 1982. 31 leaves.

Continues the story of Robot. Tells his adventures on his first day at school, and how he helped tally a survey on school cafeteria food preferences. Illustrated with color drawings.

554. Dann, Max. *Ernest Pickle's Remarkable Robot.* Illustrated by David Pearson. Melbourne: Oxford University Press, 1984. 136 p.

Tells the story of a boy, his homemade robot, and their adventures at school and when kidnapped by high-technology thieves.

555. D'Ignazio, Fred D. *Chip Mitchell: The Case of the Robot Warriors*. Illustrated by Larry Pearson. New York: E. P. Dutton, 1984. 119 p.

Chronicles the adventures of a boy, his robot, and his friends, all computer enthusiasts, and the mysteries and puzzles they solved. Gives the reader the opportunity to guess how each mystery or puzzle was solved before reading the solutions at the end of the book. Illustrated with drawings.

556. Fettig, Art. *The Three Robots*. Illustrated by Joe Carpenter. Battle Creek, MI: Growth Unlimited, Inc., 1981. 96 p. Also available on sound cassette. See number 1505.

Relates the story of three robots, one happy, one neutral, and one sad, and how all became happy robots with positive attitudes. Illustrated with drawings.

557. Fettig, Art. *The Three Robots and the Sandstorm*. Illustrated by Joe Carpenter. Battle Creek, MI: Growth Unlimited, Inc., 1983. Also available on sound cassette. See number 1506.

Continues the story of the three robots, illustrating the teamwork it takes to overcome handicaps.

558. Fettig, Art. *The Three Robots Discover Their Pos-Abilities*. Battle Creek, MI: Growth Unlimited, Inc. Also available on sound cassette. See number 1507.

Continues the adventures of the three robots, illustrating the value of goal setting.

559. Fettig, Art. *The Three Robots Find a Grandpa*. Battle Creek, MI: Growth Unlimited, Inc. Also available on sound cassette. See number 1508.

Continues the story of the three robots, illustrating a lesson in love.

560. Fettig, Art. *The Three Robots Learn About Drugs*. Illustrated by Joe Carpenter. Battle Creek, MI: Growth Unlimited, Inc., 1987. 96 p.

Continues the story of the three robots, educating about the dangers of alcohol and drug abuse and how to avoid them. Illustrated with drawings.

561. Gezi, Kal, and Bradford, Ann. *The Mystery of the Square Footprints*. Illustrated by Mina Gow McLean. Elgin, IL: The Child's World, 1980, 32 p.

Tells the adventures of a group of young people who discover a robot at the beach. Illustrated with color drawings.

562. Hoban, Lillian, and Hoban, Phoebe. *Ready...Set...Robot!* Illustrated by Lillian Hoban. New York: Harper & Row, Publishers, Inc., 1982. 64 p.

An "I Can Read" book. Tells the story of a young robot, Sol-1, who nearly lost a race due to his lack of neatness in taking care of his things. Illustrated with color drawings.

563. Hoban, Lillian, and Hoban, Phoebe. *The Laziest Robot in Zone One*. Illustrated by Lillian Hoban. New York: Harper & Row, Publishers, Inc., 1983. 64 p.

An "I Can Read" book. Tells the story of how a young robot named Sol-1 helped others and lost his reputation of being a lazy robot. Illustrated with color drawings.

564. Keller, Charles. *Ohm on the Range*. Illustrated by Art Cumings. Englewood Cliffs, NJ: Prentice-Hall, Inc., 1982. 46 p.

A collection of robot and computer jokes and riddles. Illustrated with drawings.

565. Key, Alexander. *Sprockets - A Little Robot*. Philadelphia: The Westminster Press, 1963. 144 p.

Tells about the manufacture and adventures of a small robot, who is adopted by a family which includes a professor, his wife, and young son, helps find a purple flying saucer in Mexico, and travels to the dark side of the moon. Illustrated with drawings.

566. Key, Alexander. *Rivets and Sprockets*. Philadelphia: The Westminster Press, 1964. 160 p.

Continues the story of Sprockets and his little "brother," Rivets. Relates their adventures with the professor and his son, and the occupants of a purple flying saucer, in space and on Mars. Illustrated with drawings.

567. King, Regina. *Lion-O Walks the Plank: A Thundercats Adventure*. Illustrated by Mones. Adapted from the teleplay by William Overgard. New York: Random House, 1986. 31 leaves.

Tells the adventures of Lion-O and Mandora of Third Earth and a pirate and his robots. Illustrated with color drawings.

568. Korman, Justine. *Robotman & Friends at School*. Illustrated by John Costanza. New York & Racine, WI: Golden Press & Western Publishing Company, Inc., 1986. 23 leaves.

A "Little Golden Book." Tells the story of how robotman and his friends found love from a boy, a dog, and children at school. Illustrated with color drawings.

569. Kroll, Steven. *Otto*. Illustrated by Ned Delaney. New York: Parents Magazine Press, 1983. 39 leaves.

Tells, in verse, the story of Otto, a robot who lives with a girl and her family, and who goes to school and plays a game of chase with the children. Illustrated with color drawings.

570. Lely, James A. *Star Wars*. Mankato, MN: Creative Education, 1979. 32 p.

Describes how the movie "Star Wars" was made, and tells about some of its actors and actresses, and the robots, Artoo Detoo (R2D2) and See Threepio (C3PO). Written at an elementary reading level. Includes photographs.

571. Logan, Nora. *Rim, the Rebel Robot*. Illustrated by John O'Brien. New York: Scholastic Inc., 1984. 57 p.

A "Pick-A-Path book." Tells the story of the reader's adventures with a new robot built by the reader. The reader makes decisions on how the story should unfold. Illustrated with drawings.

572. Marshall, Edward. *Space Case.* Illustrated by James Marshall. New York: The Dial Press, 1980. 38 p.
Tells the adventures of a robot space vehicle that lands on earth on Hallowe'en and goes with a boy to his school class.

573. McEvoy, Seth. *Robo Force and the Giant Robot.* Illustrated by Penalva. New York: Random House, 1985. 32 p.
Tells an adventure story involving members of the Robo Force Team led by robot Maxx Steele, and the robot Vulgar. Illustrated with color drawings.

574. McEvoy, Seth. *Robo Force and the Mountain of Burning Ice.* Illustrated by Courtney. New York: Random House, 1985. 32 p.
Tells the adventures of robot Maxx Steele and Robo Force in an engagement with enemy robots. Includes decisions for reader to make.

575. Miller, Bob, with words by Paul, Sherry. *2-B and the Rock 'n Roll Band.* Chicago: Contemporary Perspectives, Inc., for Childrens Press, 1981. 29 leaves.
A "See How I Read" book, with sentences in verse form. Tells the story of a robot, 2-B, created by mistake, who is unappreciated until he helps a band sing its way to fame. Illustrated with color drawings.

576. Miller, Bob, with words by Paul, Sherry. *2-B and the Space Visitor.* Chicago: Contemporary Perspectives, Inc., for Childrens Press, 1981. 31 leaves.
A "See How I Read" book, with sentences in verse form. Continues the story of 2-B the robot. Relates how a robot photographer from outer space reacts to a number of encounters with objects and living beings on earth, and how

the robot carries 2-B away with it. Illustrated with color drawings.

577. Milton, Joyce. *Ruthie the Robot*. Illustrated by Andrew Vines. New York: Scholastic Inc., 1983. 31 p.

Tells the story of a robot and her adventures after she leaves the robot laboratory.

578. Montgomery, R. A. *Your Very Own Robot*. New York: Bantam Books, 1982. 54 p.

A "Choose Your Own Adventure" book. The reader chooses different paths through the story to decide what to do and to decide how to cope with the adventures of a discarded robot. Illustrated with drawings.

579. Pieńkowski, Jan. *Robot*. New York: Delacorte Press, 1981. 12 p.

All-color, pop-up book about a robot family. A manual activity is provided on almost every page. Can be used with preschool children.

580. Prince, Alison. *The Type One Super Robot*. New York: Four Winds Press, 1988. 128 p.

Tells the adventures of a boy in England and the robot that was supposed to take care of him. Illustrated with drawings.

581. Rinkoff, Barbara. *Harry's Homemade Robot*. Illustrated by Leonard Shortall. New York: Crown Publishers, Inc., 1969. 38 leaves.

Relates the story of a boy's attempt to help his family with his homemade robot. Illustrated with drawings.

582. Sánchez, José Luis Garcia, and Pacheco, Miguel Angel. *A Boy and His Robot*. Illustrated by Asun Balzola. New York: Methuen, 1978. 30 leaves.

Part of a series on the rights of children sponsored by the United Nations. Originally published as *El Niño y el Robot* by Ediciones Altea, Madrid. Tells the story of a boy who was taken care of by a robot until he was put in an orphanage, and how the robot managed to visit him there. Illustrated with color drawings.

583. Seuling, Barbara. *Just Me*. New York: Harcourt Brace Jovanovich, Publishers, 1982. 64 p.
Relates the adventures of a girl who imagines she is, in turn, a horse, a dragon, and a robot. Illustrated with color drawings.

584. Slote, Alfred. *My Robot Buddy*. Illustrated by Joel Schick. Philadelphia & New York: J. B. Lippincott Co., 1975. 92 p.
Tells the story of a ten-year-old boy of the future who goes to a factory on his birthday and has a boy robot companion made. Their fun together and how they tricked a robotknapper who could not decide which one of them was a robot make up the rest of the tale. Illustrated with drawings.

585. Slote, Alfred. *C. O. L. A. R.* Illustrated by Anthony Kramer. New York: Avon Books, 1983. 146 p. Original edition published by J. B. Lippincott, 1981.
Continues the story of the robot "brother" of a ten-year-old boy, introduced in *My Robot Buddy*. Describes their voyage to, and discoveries on, a planet inhabited by runaway robots built by the factory where the boy's robot was built. Illustrated with drawings.

586. Slote, Alfred. *Omega Station*. Illustrated by Anthony Kramer. New York: J. B. Lippincott, 1983. 147 p.
Continues the adventures of a ten-year-old boy and his robot twin on the planet C. O. L. A. R. and the space ship *Omega Station*. Illustrated with drawings.

587. Steneman, Shep, adaptor. *The Black Hole Storybook*. New York: Random House, Inc., 1979. 61 p.
Based on the movie *The Black Hole*, tells the story of the crew, including the robot V.I.N.C.E.N.T., of the spaceship U. S. S. *Palomino*, and its encounter with the human captain and

humanoid and robot crew of the spaceship U. S. S. *Cygnus*, near the Black Hole. Illustrated with color photographs.

588. Titus, Eve. *Anatole and the Robot*. Illustrated by Paul Galdone. New York: Whittlesey House, 1960. 32 p.

Relates the adventures of Anatole the mouse in a cheese factory guarded by a robot. Illustrated with color and black-and-white drawings.

589. Tweedt, Joy Ann; Tweedt, Craig L.; and Granowsky, Alvin. *Robin Drake, a Robot*. Illustrated by Michael L. Denman. Cleveland, OH: Modern Curriculum Press, 1986. 31 p.

Tells the story of a family's robot helper and what happened when she fell and broke some parts. Illustrated with color drawings.

590. Waddell, Martin. *Harriet and the Robot*. Illustrated by Mark Burgess. Boston: Little, Brown and Company, 1987. 79 p.

Relates the story of a girl, the girl robot she made for her friend's birthday, and what happened when the three of them went to school. Illustrated with drawings.

591. Wilkes, Marilyn Z. *C. L. U. T. Z.* Illustrated by Larry Ross. New York: The Dial Press, 1982. 120 p.

Tells the story of C. L. U. T. Z., a robot hired by a family to do their domestic chores, and the adventures he had while working there. Illustrated with drawings.

592. Wilkes, Marilyn Z. *C. L. U. T. Z. and the Fission Formula*. Illustrated by Larry Ross. New York: Dial Books for Young Readers, 1985. 136 p.

Continues the adventures of a boy and his robot, C. L. U. T. Z., in which they try to unravel mysterious goings-on at a soda pop factory. Illustrated with drawings.

593. Williams, Jay, and Abrashkin, Raymond. *Danny Dunn, Scientific Detective*. Illustrated by Paul Sagsoorian. New York: McGraw-Hill Book Company, 1975. 172 p.

Relates the adventures of a boy detective and his robot bloodhound as they try to find the missing manager of a department store. Illustrated with drawings.

594. Woolley, Catherine (Thayer, Jane). *Applebaums Have a Robot!* Illustrated by Bari Weissman. New York: William Morrow & Company, 1980. 30 leaves.

Tells the story of Doughy, a little robot who worked in a bakery, and what happened when he worked too hard with his six hands. Illustrated with color drawings.

595. Yolen, Jane. *The Robot and Rebecca; The Mystery of the Code-Carrying Kids*. Illustrated by Jurg Obrist. New York: Alfred A. Knopf, 1980. 89 p.

Describes the adventures of a little girl and the robot she received for her ninth birthday in the year 2121. After naming it Watson II, she programs it as a detective and the two of them solve a mystery involving two thought-to-be children.

596. Yolen, Jane. *The Robot and Rebecca and the Missing Owser.* Illustrated by Lady McCrady. New York: Alfred A. Knopf, 1981. 88 p.

Continues the story of a girl and her robot Watson II as they try to first determine what an owser is, and then who took it. Illustrated with drawings.

Secondary Grades Level

Nonfiction

597. Meyer, Walter. "Robot Geometry and the High School Curriculum." *The Computing Teacher* 15 (April 1988), 23-29.

Presents eleven geometry problems and solutions that involve the movements of robots, and ideas on how to use the problems in teaching geometry. Illustrated with drawings.

Fiction

598. Ashley, Mike, compiler. *Souls in Metal; An Anthology of Robot Futures*. New York & London: St. Martin's Press, Inc., & Robert Hale, 1977. 207 p.
Anthology of nine stories about robots, by Brian W. Aldiss, Isaac Asimov, L. Sprague de Camp, Lester del Rey, Philip K. Dick, Harry Harrison, Henry Kuttner, Murray Leinster, and Clifford D. Simak.

599. Asimov, Isaac. *Foundation and Earth*. Garden City, NY: Doubleday & Company, Inc., 1986. 356 p.
Tells the story of a group of humans traveling through space, during which they meet robots in various places.

600. Asimov, Isaac. *Robot Dreams*. New York: Berkley Books, 1986. 349 p.
Includes an introduction, 20 previously published robot stories by Asimov, and one never-before published story, "Robot Dreams." Illustrated with drawings.

601. Asimov, Isaac. *Robots and Empire*. Garden City, NY: Doubleday & Company, Inc., 1985. 383 p.
A story of robots and humans in the future. Introduces the Zeroth Law: "A robot may not injure humanity or, through inaction, allow humanity to come to harm."

602. Asimov, Isaac. *The Caves of Steel*. Garden City, NY: Doubleday & Company, Inc., 1954. 224 p.
Relates the adventures of a detective and his robot partner, working together to solve a murder in New York City of the future.

603. Asimov, Isaac. *The Complete Robot*. Garden City, NY: Doubleday & Company, Inc., 1982. 557 p.
Contains an introduction and 31 previously published robot stories by Asimov, written between 1939 and 1977. Includes "Robbie," published in 1940, in which the first use of the word "robotics" appeared, and "Runaround," published in

1942, in which Asimov's Three Laws of Robotics first appeared.

604. Asimov, Isaac. *The Naked Sun*. Garden City, NY: Doubleday & Company, Inc., 1957. 187 p.
Relates the adventures of a detective and his robot partner, working together to solve a murder mystery on another planet.

605. Asimov, Isaac. *The Rest of the Robots*. Garden City, NY: Doubleday & Company, Inc., 1964. 556 p.
Includes ten robot stories by Asimov, first published in various science fiction magazines between 1941 and 1957. Two stories about the laws of robotics are among the ten.

606. Asimov, Isaac. *The Robot Collection*. Garden City, NY: Doubleday & Company, Inc., 1983. 557 p.
Contains the robot novels *The Caves Steel* and *The Naked Sun* and 31 previously published robot stories by Asimov. Included are "Robbie," published in 1940, in which appears the first use of the word "robotics," and "Runaround," published in 1942, in which Asimov's Three Laws of Robotics appear.

607. Asimov, Isaac. *The Robot Novels*. New York: Ballantine Books, 1988. 671 p.
Contains the robot novels *The Caves of Steel*, *The Naked Sun*, and *The Robots of Dawn*.

608. Asimov, Isaac. *The Robots of Dawn*. Garden City, NY: Doubleday & Company, Inc., 1983. 408 p.
Relates the adventures of a detective and his robot partner, working together to solve the murder of a humaniform robot on another planet.

609. Asimov, Isaac; Warrick, Patricia S.; and Greenberg, Martin H. *Machines That Think*. New York: Holt, Rinehart and Winston, 1983. 627 p.
Contains science fiction stories about robots and computers. Authors include Paul Anderson, Isaac Asimov,

Harry Bates, Ambrose Bierce, J. F. Bone, Fredric Brown, John Brunner, Arthur C. Clarke, Lester del Rey, Philip K. Dick, Gordon R. Dickson, Harlan Ellison, Harry Harrison, Murray Leinster, William M. Miller, Jr., Michael Shaara, Robert Silverberg, A. E. van Vogt, Harl Vincent, Vernor Vinge, Robert Moore Williams, Gene Wolfe, John Wyndham, and George Zebrowski.

610. "Isaac Asimov's Robot City."
Series of science fiction novels on Asimovian robots written by various authors, with Isaac Asimov as consultant. Includes the following works, numbers 611-619.

611. Kube-McDowell, Michael P. *Book l: Odyssey*. New York: Ace Books, 1987. 211 p. Also available on sound cassette. See number 1508a.

612. McQuay, Mike. *Suspicion*. New York: Ace Books, 1987. 177 p.

613. Cover, Arthur Byron. *Prodigy*. New York: Ace Books, 1988. 176 p.

614. Wu, William F. *Pedrihelion*. New York: Ace Books, 1988. 162 p.

615. Leigh, Stephen. *Changeling*. New York: Ace Books, 1989. 151 p.

616. Scotten, Cordell. *Renegade*. New York: Ace Books, 1989. 178 p.

617. Bethke, Bruce. *Maverick*. New York: Ace Books, 1990. 184 p.

618. Oltion, Jerry. *Alliance*. New York: Ace Books, 1990. 168 p.

619. Thurston, Robert. *Intruder*. New York: Ace Books, 1990. 195 p.

620. Asimov, Janet, and Asimov, Isaac. *Norby* Series.
Series about the adventures of fifteen-year-old Jeff and robot Norby. See numbers 530-540.

621. Bellairs, John. *The Eyes of the Killer Robot*. New York: Dial Books for Young Readers, 1986. 167 p.
Relates the story of a 13-year-old boy, kidnapped by the owner of a killer robot to obtain a pair of nearsighted eyes.

622. Binder, Eando. *Adam Link - Robot*. New York: Paperback Library, 1965. 174 p.
The story of how a robot, through consistent efforts, battling criminal and alien forces, became a citizen of the United States, as told by the robot. Includes some violence.

623. Bly, Robert W. *Ronald's Dumb Computer*. Illustrated by Hodges Soileau. Wayne, PA: Miles Standish Press, 1983. 79 p.
The story of two teenagers and a robot they found. The object of the story is to show how computers work and how computer programmers use logic to make them work. Illustrated with drawings.

624. Conklin, Groff, ed. *Science Fiction Thinking Machines: Robots, Androids, Computers*. New York: The Vanguard Press, Inc., 1954. 367 p.
An anthology of 14 stories about robots, four about androids, and four about computers. Robot story authors include Isaac Asimov ("Robbie"), Ambrose Bierce, Alan Bloch, Karel Čapek (*R. U. R.*), Raymond Z. Gallun, Herbert Goldstone, Fritz Leiber, Eric Frank Russell, Michael Shaara, Clifford Simak, William Tenn, and S. Fowler Wright. Android story authors include Chan Davis, Wallace MacFarlane, Theodore Sturgeon, and Wallace West. Computer story authors include Poul Anderson, Hal Clement, Walter M. Miller, Jr., and Robert Sherman Townes.

625. Coville, Bruce. *Robot Trouble*. New York: New American Library, 1986. 221 p.

Relates the adventures of six young geniuses, the A. I. (Artificial Intelligence) Gang, on an island off the coast of the Carolinas, and their plan to launch a musical robot into space.

626. del Rey, Lester. *Robots and Changelings*. New York: Ballantine Books, 1957. 175 p.
A collection of 11 stories about robots.

627. del Rey, Lester. *The Runaway Robot*. New York: Westminister Press, 1965. 188 p. Also published by Scholastic Book Services, New York, 1965.
Relates the adventures of 16-year-old Paul and his robot Rex in space, as told by the robot.

628. Dick, Philip K. *Robots, Androids, and Mechanical Oddities: The Science Fiction of Philip K. Dick*. Carbondale: Southern Illinois University Press, 1984. 261 p.
A collection of surprise-ending science fiction stories.

629. D'Ignazio, Fred. *Robot Odyssey 1: Escape from Robotropolis*. New York: Tor Books, 1988. 284 p.
Tells the story of a young inventor of robots and his adventures in Robotropolis, based partly on *Robot Odyssey*, computer learning game developed, marketed, and owned by The Learning Company, Menlo Park, California. See number 1485.

630. Elwood, Roger, ed. *Invasion of the Robots*. New York: Paperback Library, Inc., 1965. 157 p.
An anthology of eight stories about robots by Isaac Asimov, Robert Bloch, Lester del Rey, Philip K. Dick, Henry Kuttner, Richard Matheson, Eric Frank Russell, and Jack Williamson.

631. Fairman, Paul W. *The Forgetful Robot*. New York: Holt, Rinehart & Winston, 1968. 163 p.
Tells the story of a near-human robot, from his perspective, as he tries to help an actor family, journeys with them to New Mars, and becomes a part of plans of a zealous

scientist who is hoping to arm the hidden and forgotten inhabitants of an abandoned city.

632. Greenberg, Martin. *The Robot and the Man*. New York: The Gnome Press, Inc., 1953. 251 p.
An anthology of ten stories about robots, including stories by John S. Browning, Lester del Rey, H. H. Holmes, Joseph E. Kelleam, John D. MacDonald, Lewis Padgett, A. E. van Vogt, Robert Moore Williams, and Bernard Wolfe.

633. Harrison, Harry. *The Men from P. I. G. and R. O. B. O. T.* New York: Atheneum Publications, 1978. 141 p.
Tells the stories of how two men, one with a herd of pigs and one with robots, solved mysteries on the planet Trowbri.

634. Harrison, Harry. *War with the Robots*. London: Dennis Dobson, 1962. 158 p.
A collection of eight science fiction stories about robots.

635. Hoover, H. M. *Orvis*. New York: Viking Kestrel, 1987. 186 p.
Tells the adventures of a girl and an obsolete robot on an inhospitable Earth of the future.

636. Kandel, Michael. "Stanislaw Lem on Men and Robots." *Extrapolation* 14 (December 1972), 13-24.
Focuses upon the science fiction works of Stanislaw Lem, such as *Tales of Pirx the Pilot* and *Robot Fables*, which involve robots. Also discusses *I, Robot* by Isaac Asimov.

637. Kraft, David Anthony, with programming by Zakar, Susan M. *Robot Race*. New York: Scholastic Inc., 1984. 26 p.
"Micro Adventure #6." An adventure story in which the reader uses computer programs provided throughout the text to outwit a half-human, half-computer evil robot who is leading a robot army against the world.

638. Leroe, Ellen W. *Robot Romance*. New York: Harper & Row, Publishers, 1985. 179 p.

Tells the story of a 16-year-old student at robot-run Silicon Computer High School, whose student body is made up of robots and humans, and the girl humanoid robot he built for the science fair.

639. Leroe, Ellen W. *Robot Raiders*. New York: Harper & Row, Publishers, 1987. 181 p.

A continuation of *Robot Romance*. Tells how a 16-year-old student at the Massachusetts Institute of Technology works with the National Aeronautics and Space Administration to reconstruct his girl humanoid robot for a Mars landing project, and the adventures he encounters.

640. Lewis, Arthur O., Jr. *Of Men and Machines*. New York: E. P. Dutton & Co., Inc., 1963. 349 p.

An anthology of 55 fiction and nonfiction stories and writings about the relationship of humans and machines. Included are selected writings on robots, automatic objects, and machinery, arranged under the following categories: The Problem; The Machine as Friend; The Machine as a Thing of Beauty; The Machine as Enemy; and Futures. The earliest selection included is "The House of Salomon," written by Sir Francis Bacon in 1624. Other early authors included are Adam Smith (1776), James Reynolds (1802), Samuel Miller (1803), George Gordon, Lord Byron (1812), and Edgar Allan Poe (1829). Also in the anthology are articles, stories, and excerpts of works by Brian W. Aldiss, Sherwood Anderson, Isaac Asimov, W. H. Auden, Steven Vincent Benet, Margot Bennett, Morris Bishop, Macknight Black, Ray Bradbury, Samuel Butler, Karel Čapek, Walter van Tilburg Clark, Stephen Crane, e e cummings, Emily Dickinson, Ralph Waldo Emerson, Paul Engle, E. M. Forster, Robert Frost, Wolcott Gibbs, Joe Glazer, Aldous Huxley, Donald N. Michael, Lewis Mumford, George Orwell, George R. Price, Carl Sandburg, Adam Smith, C. P. Snow, Stephen Spender, John Steinbeck, George Strauss, Harvey Swados, Robert Sherman Townes, Mark Twain, Kurt Vonnegut, Jr., Thomas J. Watson, Jr., Edward C. Weir, E. B. White, Walt Whitman, and Reed Whittemore.

641. Macrae, G. V. *G. I. Joe: Operation: Robot Assassin*. New York: Ballantine Books, 1985. 90 p.

A "Find Your Fate" book. Relates the adventures of a bomb specialist on a G. I. Joe squad. Reader makes decisions that determine the plot. Illustrated with drawings.

642. Martin, David. *K9 and the Zeta Rescue*. Windermere, FL: Rourke Publications, Inc., 1982. 31 p.

A Dr. Who book. Tells the adventures of K9, a robot dog who rescues survivors of an interplanetary war. Illustrated with color drawings.

643. McKinney, Jack. *Genesis*. New York: Ballantine Books, 1987. 214 p.

Volume 1 of 18-volume series by this author, based on animated "Robotech" series, a saga of people and robots in the future. See numbers 1148, 1149, 1151, and 1152 for information on how the animated series was produced, number 650 for a game based on the series, and number 644 for the last book title in McKinney's series.

644. McKinney, Jack. *The End of the Circle*. New York: Ballantine Books, 1990. 343 p.

Volume 18 of 18-volume series by this author, based on animated "Robotech" series, a saga of people and robots in the future. See numbers 1148, 1149, 1151, and 1152 for information on how the animated series was produced, number 650 for a game based on the series, and number 643 for the first book title in McKinney's series.

645. Meddick, Jim. *Robotman: The Untold Story*. New York: Topper Books, 1986. 128 p.

A collection of comic strips featuring Robotman, a family robot.

646. Moskowitz, Sam. *The Coming of the Robots*. New York: Collier Books, 1963. 254 p.

An anthology of ten stories about robots by Isaac Asimov, Eando Binder, Lester del Rey, Michael Fischer,

Raymond Z. Gallun, Peter Phillips, Clifford D. Simak, F. Orlin Tremaine, Harl Vincent, and John Wyndham.

647. Nolan, William F. *The Pseudo-People; Androids in Science Fiction*. Los Angeles, CA: Sherbourne Press, 1965. 238 p.

An anthology of 14 stories about robots by Frank Anmar, Isaac Asimov, Charles Beaumont, Ray Bradbury, James Causey, Dennis Etchison, Charles E. Fritch, Ron Goulart, Henry Kuttner, Shelly Lowenkopf, Richard Matheson, William F. Nolan, Chad Oliver, and Robert F. Young. Includes an introduction by A. E. van Vogt.

648. Petit, Jean-Pierre. *Run, Robot, Run; The Adventures of Archibald Higgins*. Translated by Ian Stewart. Los Altos, CA: William Kaufmann, Inc., 1982. 72 p.

Cartoon story in which a young turtle robot inventor's experiments with robots and other devices introduce the concepts of robotics and artificial intelligence. Covers programmable automata, input-output systems, signal analysis, response time, servomechanisms, control and homeostatic systems, pattern recognition, noise, image processing, perception, trial-and-error strategy, the Turing Test, and artificial intelligence.

649. Saberhagen, Fred. *Berserker*. New York: Ballantine Books, 1967. 190 p.

A science fiction story about a machine set up by people on earth to destroy anything that had life.

650. Siembieda, Kevin. *Robotech: The Role Playing Game*. Detroit: Palladium Books, 1988. 110 p.

Examines the characters, planes, equipment, and weapons of "Robotech," the animated film and television series, and describes a role playing game using the information given. See numbers 1148, 1149, 1151, and 1152 for information on how the animated series was produced, and numbers 643 and 644 for information on an 18-title series of books by Jack McKinney based on the "Robotech" animated series.

651. Silverberg, Robert. *Men and Machines*. New York: Meredith Press, 1968. 240 p.
An anthology of ten stories of science fiction, including several about robots. Authors include Brian W. Aldiss, James Blish, Lester del Rey, Randall Garrett, Fritz Leiber, Lewis Padgett, Fred Saberhagen, Robert Silverberg, George O. Smith, and Jack Williamson.

652. Simak, Clifford D. *City*. Garden City, NY: Nelson Doubleday & Company, Inc., 1952. 211 p.
Tells the story of an Earth run by dogs and robots after most humans have moved to Jupiter.

SECTION 4. OTHER EDUCATIONALLY USEFUL ROBOTS

This section includes materials on personal, hobby, and toy robots whose focus is not on educational uses but on the robot as a robot. However, information on many of the robots covered can also be found in the Educational Robots grouping.

Personal robots lend themselves to numerous applications in the school and classroom as well as in the home. A robot that can fetch a drink from the kitchen might also be able to fetch papers from students or deliver items to the principal's office. A robot designed to serve as a home security system or as a gas, fire, and smoke warning device perhaps could perform these same functions in the school corridors. Teachers and school administrators will find many uses for personal robots as they become more sophisticated. All personal robots covered in the listed materials could be used for robotics demonstrations in the classroom.

For this sourcebook, a hobby robot is considered to be a robot that is built by the hobbyist, either from a kit or from parts, using plans or the hobbyist's knowledge and imagination. Materials listed supply many plans that are well adapted to the classroom for use as a group or individual student activity or project. Kits are listed in Appendix D under Educationally Useful Robots. Appendix E provides addresses of robot parts and electronics suppliers. Local electronics supply stores are also good sources of needed items.

Toy robots include transformable robot toys, windup and mechanical robot toys, and other robots which can do no obvious work but are meant to simply entertain. These robots are excellent for demonstrating robotics principles to preschool and elementary grades level students, and as props when telling stories about robots. They also serve well in displays and class projects showing what robots can do. Toy robots combined with toy human figures can be used to illustrate and motivate discussions about the social, workplace, economic, and physical (e.g., safety) impacts of robots. Students' personal collections of toy robots can be displayed, creating a springboard for further discussions.

The materials listed in the Toy Robots grouping show the wide variety available and the great interest in collecting these toys.

Personal Robots

653. "A Home-Made 'Otto-Tron': The Ideal Pal." *Ebony* 25 (June 1970), 64-66, 68.

Describes a seven-foot robot built by James Chisholm of Portsmouth, Virginia. Illustrated with photographs.

654. "Advanced Home Robot Rivals Factory Kin." *Machine Design* 58 (February 6, 1986), 38-39.

Describes Hero 2000, with some comparison with Hero 1. Includes color photographs and detailed drawings.

655. "Do-It-Yourself Robot Has On-board Computer." *Design News* 39 (January 10, 1983), 14.

Describes Hero 1. Illustrated with photograph.

656. "Home Robot Literally Is a Handyman." *Design News* 39 (February 7, 1983), 26, 28.

Describes Jenus, robot that greets guests, serves refreshments, entertains, patrols premises, detects smoke, carbon gas, and gamma radiation, and vacuums carpets, made by World of Robots Corporation, Jackson, Michigan. Illustrated with photograph.

657. "Household Robot." *Popular Science* 236 (May 1990), 13.

Reports on Droid Personal Robot I, developed by Droid Systems Corporation, 412 East 88th Street, New York, NY 10128. Illustrated with color photograph.

658. "Nolan Bushnell: Automated au Pairs." *Economist* 285 (October 30, 1982), 66-67.

Describes Bushnell's plans for new ventures, Pizza Time Theatre (restaurant chain using robot entertainment), and Androbot, Incorporated, developer of personal robots. Illustrated with photograph.

659. "Personal Robot." *Popular Science* 235 (November 1989), 104.

Illustrates, with color photograph, Newton, personal robot from SynPet Personal Electronic Technologies, 7225 Franklin Road, Boise, ID 83709.

660. "Personal Robotics." *Popular Mechanics* 166 (October 1989), 45.

Reports on two personal robots, Newton, produced by SynPet Personal Electronic Technologies, Boise, Idaho, and a robot produced by Droid Systems Corporation (see number 657).

661. "Personal Robots Characteristics and Sources." *Computers & Electronics* 22 (December 1984), 88-90.

Supplies information and addresses for 21 personal robots and robot families. Includes Alpha I/II, Armadillo, Armatrol, Armatron, Armborg, Armdraulic, Armover, Armsort, B. O. B., F. R. E. D., Genus, Hero 1, Hero Jr., HuBot, Itsa-Box, Marvin, Maxx Steele, Memocon Crawler, MiniMover, Movemaster, RB5X, Rhino, Robot I, Robot IV, Scorpion, Tasman Turtle, TeachMover, Topo, Turtle Tot, and Zeaker.

662. "The Robots Are Coming." *Business Week* no. 2769 (December 13, 1982), 62-63.

Describes the state of the personal robot market at that time. Mentions Hero, RB5X, Topo, and Genus 1. Includes photographs.

663. "The Wrights and Their Robot 'Kids.'" *Ebony* (March 1987), 72, 74, 76.

Reports on robots built by Dee and Sam Wright of Baltimore, Maryland, used in their presentations of how robots can be used with the elderly and as companions. Includes color photographs.

664. "Will the Robot Be Father to the Industry?" *Inc.* 5 (March 1983), 64.
Reports on robots Topo and B. O. B. Illustrated with photograph.

665. A Kind Human Being. "January Robotmate of the Month: Hero 1." *Computers & Electronics* 21 (January 1983), 46-47, 94, 96.
Describes Hero 1. Includes photographs.

666. Angier, Natalie. "Fun and Learning with Science Toys." *Discover* 2 (December 1981), 46-51.
Includes description and photograph of ComRo and robot pet Wires, and mention and photograph of fischer america construction kit which includes robot.

667. Aronson, Robert B. "High-Tech Recreation Equipment." *Machine Design* 57 (April 25, 1985), 50-56.
Includes information and photographs for Omnibot 2000 and Hero Jr.

668. Baker, Leslie, and Sakowski, Nat. "The Robots Are Coming. Why Not Introduce One to Your Apple?" *A+* 2 (April 1984), 72-74, 76-79.
Discusses personal robots, including B. O. B., F. R. E. D., RB5X, and Topo, ways to interface these robots to an Apple II+ or IIe computer, and what they can do. Also reports on advantages and applications of robots in general. Boxes explain what a robot is, and report on a New York City robot used to paint sanitation trucks. Illustrated with color drawing.

669. Bateman, Selby. "Robots: The New Mobile Computer." *Compute's Gazette* 2 (April 1984), 26, 30, 32, 34, 36, 38.

Discusses the personal robot market, obstacles in robot development, discriminating vision, bipedal locomotion, an adaptable hand, and a natural language, and describes Topo, F. R. E. D., B. O. B., Hero 1, and RB5X. Illustrated with photographs.

670. Bernard, Josef. "Hero Jr." *Computers & Electronics* 22 (December 1984), 63-64.
Describes Hero Jr. Includes photographs.

671. Blankenhorn, Dana. "Personal Robots: The PC on Wheels." *PC World* 3 (January 1985), 177-184.
Discusses personal robots, including RB5X, B. O. B., and Hero 1, programming robots, robot legs and feet, arms and hands, mouths and ears, balance and touch, eyes, brains, and Isaac Asimov's Three Laws of Robotics. Illustrated with drawings and color photographs.

672. Boyer, Howard. "Robots: The New Helpmates?" *Working Woman* (April 1984), 134, 136.
Considers how present robots can help in the home. Describes specifically B. O. B., Genus, Hero 1, and RB5X. Cover of this issue shows color photograph of RB5X.

673. Cavaliere, Frank. "Close Encounter Illustrated Robot Market's Faults, Future. While Entertaining, Today's Robots Still Offer Little Utility." *Merchandising* 9 (December 1984), 31, 33.
Describes Hero Jr. Illustrated with photograph.

674. Chase, Victor D. "Open House with a Live-In Robot." *Popular Mechanics* 150 (October 1978), 200, 202.
Describes Arok, a personal robot built by Ben Skora of Palos Hills, Illinois, and ways it helps around the house and at promotions. Includes photographs.

675. Coll, Steve. "When the Magic Goes." *Inc.* 6 (October 1984), 82-84, 90, 92, 94, 97.

Relates the story of Nolan Bushnell, founder of Atari and Pizza Time Theatres (restaurants using robots for entertainment), and a founder of Androbot, Incorporated, maker of B. O. B., F. R. E. D., and Topo. Illustrated with photographs. Cover of this issue shows a photograph of Nolan Bushnell.

676. Conway, John. "Personal Robots; New Products Offer a Chance to Get Hands-on Experience with Robotics for a Relatively Modest Cost." *Computers & Electronics* 22 (December 1984), 60-63.
Discusses types of personal robots (turtles, arms, mobiles), who will buy them, what they can and cannot do, and the future. Includes photographs.

677. Cornish, Edward. "The Home of Tomorrow." *Consumer's Digest* 24 (March/April 1985), 78-81.
Includes description of ways robots may be used in the home.

678. D'Ignazio, Fred. "The Robot Teddy Bear." *Compute!* 6 (January 1984), 94-104.
Describes how D'Ignazio's family uses personal robots.

679. Faflick, Philip. "Here Come the Robots." *Time* 121 (March 7, 1983), 76-77.
Reports on the personal robot market. Illustrated with photographs.

680. Fox, Steve. "Cutebots." *Omni* 8 (April 1986), 22, 113.
Describes Omnibot 2000, Elami Jr., Hearoid, VideoRobot (Nintendo), the Capsela 5000 construction kit, the Movits kits, Es-Car-Go, and Wise Argent Orb (WAO). Illustrated with color photographs.

681. Gorzelany, Jim. "They Walk They Talk But..." *Consumer's Digest* 24 (July/August 1985), 35.
Describes Omnibot, Hero Jr., and HuBot, and discusses the use of personal robots. Illustrated with drawing.

682. Govoni, Steve. "Raves for Slaves." *Financial World* 155 (April 29, 1986), 10.

Reports on Joseph F. Engelberger and his company, Transitions Research Corporation, of Danbury, Connecticut, which aims to develop a robot capable of fixing a broken dishwasher, making dinner, and clearing the table. Illustrated with photograph.

683. Greenwalt, Julie. "Man's New Best Friend Is a Walking, Talking, Singing Robot-in-a-Kit Named Hero." *People Weekly* 19 (March 28, 1983), 63-64, 66.

Reports on an interview with Hero 1, helped out by inventor Jim Lytle and Douglas Bonham, director of educational products for Heath, manufacturer of Hero. Describes Hero and what the robot can do. Includes photographs.

684. Hall, Alan. "The Grass Needs Cutting? Here Comes the 'Homebot.'" *Business Week* no. 3042 (March 14, 1988), 60.

Reports on study to develop a robot that can do household chores and be a geriatric companion, being conducted by Joseph F. Engelberger's Transitions Research Corporation, Danbury, Connecticut. Illustrated with color drawing.

685. Hawkins, William J. "Electronics Newsfront." *Popular Science* 228 (February 1986), 28, 30.

First part of article describes Hero 2000. Includes photographs.

686. Hero. "The New Wave of Personal Robots." *Computers & Electronics* 21 (January 1983), 37-41.

Describes Tasman Turtle, ITSABOX, MiniMover-5, TeachMover, Rhino XR-1, RB5X, and Hero 1. Provides photographs of each.

687. Higgins, Mike. *A Robot in Every Home*. Oakland, CA: Kensington Publishing Company, 1985. 191 p.

Covers personal robots, past, present, and future, and the following personal robots: Topo, Hero 1, RB5X, HuBot, Hero Jr., Turtle Tot, F. R. E. D., Dingbot, Verbot, and Omnibot. Also provides information on personal robot accessories, manufacturers, retailers, and distributors, and periodicals and associations, and a glossary and bibliography. Illustrated with photographs and drawings.

688. Higgins, Mike. "The Future of Personal Robots." *The Futurist* 20 (May/June 1986), 43-46.

Reports on ways personal robots are being used, discusses the sociology of robots, and describes the "super-robot" and the self-aware robot. Considers philosophical and ethical questions involved. Includes photographs of personal robot uses. Excerpted from *A Robot in Every Home: An Introduction to Personal Robots & Brand-Name Buyer's Guide*.

689. Higgins, Mike. "Personal Robots Face Software Challenge." *High Technology* 5 (March 1985), 71-73.

Examines the problem of software for personal robots. Describes software accompanying the following robots: Topo, RB5X, Hero 1, and Hero Jr. Includes photographs.

690. Inversen, Wesley R. "Smart Robots Find a Home." *Electronics* 56 (January 27, 1983), 79-81.

Discusses personal robots, especially Hero 1, Topo, and B. O. B. Illustrated with photographs.

691. Kahn, Joseph. "Death of a Robot." *Inc.* 7 (February 1985), 21.

Reports on the home robot industry. Illustrated with drawing.

692. Knight, Timothy O. *Probots and People: The Age of the Personal Robot*. New York: McGraw-Hill Book Company, 1984. 126 p.

Discusses the "probot" or personal robot. Considers whether humans are ready for robots, industrial robots, the practical probot, and probots of the future. Describes AndroMan, F. R. E. D., Topo, Hero 1, B. O. B., and RB5X. Illustrated with photographs and drawings.

693. Koblenz, Jay. "In Search of Personal Robots." *Black Enterprise* 17 (October 1986), 35.
Reports on Hero 2000.

694. Levy, Steven. "The Wisdom of Nolan 'King Pong' Bushnell." *Rolling Stone* no. 405 (September 29, 1983), 107.
Reports on a press conference given by Nolan Bushnell concerning personal robots. Includes photograph.

695. Lonergan, Tom, and Frederick, Carl. *The VOR*. Rochelle Park, NJ: Hayden Book Company, Inc., 1983. 120 p.
Describes the design, construction, and capabilities of OMNIVAC, the first Volitionally Operant Robot. Referred to as The VOR, it is an intelligent robot aware of its surroundings, with hearing and speech abilities. Illustrated with drawings.

696. Mamis, Robert A. "The Pied Piper of Sunnyvale." *Inc.* 5 (March 1983), 57-58, 62, 64, 66.
Relates the business venture history of Nolan Bushnell, including the robot manufacturer Androbot, Incorporated. Illustrated with photographs.

697. Marbach, William D., and Abramson, Pamela. "From Atari to Androbot." *Newsweek* 100 (November 15, 1982), 123.
Reports on business ventures of Nolan Bushnell, including Androbot, Incorporated, manufacturer of the personal robot Androbot. Includes photograph.

698. Marbach, William, with Springen, Karen. "I'm Dreaming of a Robot Hamster..." *Newsweek* 108 (December 8, 1986), 83-84.
Reports on high-technology toys and robots, including a robot hamster, Nintendo Entertainment System, Movits, and fischertechnik's robot arm. Includes color photographs.

699. Marrs, Texe. *The Personal Robot Book.* Blue Ridge Summit, PA: TAB Books, Inc., 1985. 181 p.

A very thorough coverage on the personal robot. Gives a general introduction to personal robots, and provides in-depth information, including, in some cases, ways to use specific robots in education, for the following robots: Hero 1, Hero Jr., RB5X, ComRoTot, HuBoT, Dingbot, Verbot, Omnibot, Maxx Steele, Androman, F. R. E. D., Topo III, B. O. B., RoPet, Robot Shop Kits, Servitron Robots, Marvin, ELAMI Robots, Scorpion, Armover, Armadillo, Line Tracer II, Memocon Crawler, Nomad, Smart Rabbit, Cyber I, Robomower, Robocycle, Armatron, Martha and George Hydrobot, The Turtle, Isaac, ITSABOX, and Gemini. Also provides information on robotic arm educators, and a number of other types of robots and robot kits, and custom-built robots. Includes information and addresses for parts and software for robot builders, books and magazines, robot hobbyist organizations, and manufacturers, and a glossary. Illustrated with photographs and drawings.

700. McClean, Cynthia. "Tech Watch." *Black Enterprise* 15 (June 1985), 78.

Reports on personal robot uses and market, and Hero 1, Hero Jr., and RB5X. Illustrated with photographs.

701. McComb, Gordon. "Personal Robots." *Creative Computing* 9 (November 1983), 196, 198, 200, 203-204.

Discusses personal robot applications, history of robots, and the future, and describes Topo, B. O. B., Hero, RB5X, and MiniMover 5. Includes photographs.

702. Meinert, Jim. "Robots Are Ready! Are You?" *Industrial Education* 74 (May/June 1985), 8-9.

Provides background information on industrial robots, a human - robot comparison, and description of a project involving Hero 1. Describes how Hero was taught to act as a smoke detector and fire alarm and to recognize colors.

703. Myers, Edith. "Now It's Personal Robots." *Datamation* 30 (August 15, 1984), 57-58.

Reports on the market for personal robots.

704. Neimark, Jill. "Debbie Huglin and Her Robotorium." *Ms.* 13 (September 1984), 26.

Describes Debbie Huglin's store, Robotorium, in New York City, which sells only robots. Illustrated with drawing.

705. Owens, Peter. "At Home with...the Hero 1." *Popular Computing* 3 (December 1983), 166-168, 170.

Reports on a two-week home visit by a Hero l. Presents the following principles of robot ownership: (1) An owner wants the robot to reflect the owner's own behavior. (2) A robot appears to us as essentially a social machine whose psychological reality seems to grow and develop as people react to it. (3) Our relationship with robots is essentially nondemocratic, and we are pleased when we find ways to bring them under our control. (4) We are capable of empathy and even embarrassment on behalf of the robot. (5) Household robots are inevitably imbued with human personalities. (6) Once a robot is given a certain degree of autonomy, e. g., the ability to report on its declining batteries, it gains a quantum leap in identity. (7) We respond to whole anthropomorphic entities and not to their parts. When taken apart, the robot is quickly relegated to machine status.

706. Owens, Peter. "The Hero 1." *Popular Computing* 3 (December 1983), 170-175.

Provides in-depth description of Hero 1. Includes photographs and insert on talking to Hero.

707. Peterson, Ivars. "Run, Robot, Run." *Science News* 127 (February 16, 1985), 109.

Reports on personal robots at the Consumer Electronics Show in Las Vegas, Nevada, in January 1985 and the personal robot field, and describes *ChipWits*, a computer program in which the user programs a robot to cope with various hazards that appear on the screen. See number 1495 for information on *ChipWits*. Cover of this issue features drawing of pet robot.

708. Prince, Suzan D. "Automatalk." *Creative Computing* 8 (December 1982), 156, 158, 160, 162, 164.

Reports on progress and problems in the development of personal robots, and discusses a number of robots being developed or available. Illustrated with photographs.

709. Raidt, Rose. "The Robots Are Coming." *Working Woman* 8 (August 1983), 61.

Reports on Debbie Huglin's robotics store in New York City, Robotorium, which sells only robots.

710. Resnick, Mitchel. "Robots Get Personal." *Science '83* (April 1983), 75-76.

Discusses the personal robot B. O. B., its uses and its future. Illustrated with drawing.

711. Resnick, Mitchel. "The Robots Are Coming." *Rolling Stone* no. 404 (September 15, 1983), 79-83, 85.

Reports on the personal and hobby robot field, and robots built by Androbot (Topo and B. O. B.), Gene Oldfield who owns Robot Repair in Sacramento, California, Bill Bakaleinikoff of Petaluma, California and his company, Superior Robotics, and Jim Hill of Los Angeles, California. Includes color and black-and-white photographs.

712. Robillard, Mark J. "Buyer's Guide to Personal Robots." *Radio-Electronics* 57 (March 1986), 41-46.

Discusses what a personal robot is, provides a recent history of personal robots, and describes a number of personal robots available, including Rhino robots, Alpha arms, Nomad, Gemini, the Hero family of robots, the Tomy family of robots, RB5X, the fischertechnik Robot Computing kit, and the Nintendo Entertainment System. Includes photographs. Cover of this issue shows a Gemini robot.

713. Robillard, Mark J. *Hero 1: Advanced Programming and Interfacing*. Indianapolis: Howard W. Sams & Co., Inc., 1983. 234 p.

Covers functions and programming of the 6808 family of microprocessors, the Hero 1 robot interpreter and robot language, firmware and RAM use, software enhancement, and hardware enhancement. An appendix provides definitions of the executable instructions. Illustrated with photographs, drawings, and schematics.

714. Robillard, Mark J. "Pet Robots?" *Radio-Electronics* 57 (June 1986), 74.

Describes two pet robots made by Axlon, Petster and S. O. S. Bear.

715. Robillard, Mark J. "Rovers Revisited." *Radio-Electronics* 57 (October 1986), 82-84, 91.

Describes Ernie, a programmable turtle robot, and ways it can be modified. Includes photograph.

716. Roessing, Walter. "The Robots Are Coming!" *Saturday Evening Post* 255 (November 1983), 66-67, 100-101.

Reports on robots and developments in the personal and promotional robot fields. Includes B. O. B., DC-2 (Android Amusement Company), OPD2 (Orlando, Florida Police Department educational robot), Robot Redford, a "demonstration android" used at state fairs and trade shows, owned by Bill Bakaleinikoff of Superior Robotics of America, and SICO, a robot used on NBC-TV soap opera "Days of Our Lives," from International Robotics of New York City. Also mentioned and discussed are Hero 1, ComRoTot, and RB5X. Includes color photographs.

717. Ruby, Daniel J. "Computerized Personal Robots: They Move, Talk, Think, and Teach You Robotics." *Popular Science* 222 (May 1983), 98-100, 136.

Reviews developments in the personal robotics field. Discusses capabilities of a number of personal robots, and future possibilities. Illustrated with color and black-and-white photographs. Cover of this issue shows a photograph of a Hero 1 in motion.

718. Shapiro, Neil. "30 Days with Hero-1; The New Heath Educational Robot One Turns Out To Be a Personable Member of the Family." *Popular Mechanics* 160 (September 1983), 74-75, 141.
Describes Hero 1. Includes color photographs and drawing.

719. Stuckey, Scott. "The Robot Invasion." *Boy's Life* 74 (December 1984), 30-33, 78.
Discusses what robots are and what they can do, and describes a number of personal robots and robot toys. Includes color drawing and photographs.

720. Teschler, Leland. "The First Wave of Home Robots." *Machine Design* 55 (March 24, 1983), 85-90.
Discusses personal robots, especially Hero, B. O. B., RB5X, and Genus, and problems of design. Illustrated with photographs and drawings.

721. Whalen, Bernice. "Upscale Consumers Adopt Home Robots But Widespread Lifestyle Impact Is Years Away." *Marketing News* (November 25, 1983), 5.
Discusses present and future uses of personal robots.

722. Winkless, Nelson B., III. *If I Had a Robot...What to Expect from the Personal Robot*. Beaverton, OR: dilithium Press, 1984. 247 p.
Discusses how the personal robot might be used. Considers the robot as thinker, communication and control, sensing and remembering, power sources, mechanics, and laws affecting robots. Illustrated with drawings and diagrams.

723. Wolkomir, Richard. "Robots at Home." *Omni* 5 (April 1983), 70-72, 74-76.
Considers ways personal robots may be used in the future, and reports on personal robot advocates and projects.

Illustrated with portrait of lady robot. Cover of this issue shows human hand touching robot hand, in color.

724. Young, Jeffrey S. "Macintosh Robotics." *Macworld* 3 (August 1986), 55-56.

Describes Elfin, a robot produced by Elfin Technologies, controlled by a radio-frequency generator which sets on top of the Macintosh microcomputer. Includes photograph.

Hobby Robots, Robot Building

725. "Move Over, R2-D2." *Ebony* 35 (July 1980), 56, 58.

Describes a 5-foot, 10-inch tall robot built by Gregory Thomas of East St. Louis, Illinois, to help finance his college education. Includes photographs of the robot outdoors and helping around the house.

726. "Your Own Robot? Do It Yourself, Says Wizard, 16." *Science Digest* 83 (June 1978), 20-21.

Reports on activities of 16-year-old Jonathan Kaplan and the robot he built. Includes photograph.

727. Allen, Bill. "Robots You Can Build Today." *Popular Mechanics* 154 (August 1980), 16-17, 124-125.

Describes plans and kits available for building hobby robots and provides addresses. Box reports on the Amazing Micro-Mouse Maze Contest sponsored by the Institute of Electrical and Electronics Engineers' *Spectrum* magazine. Includes color and black-and-white photographs.

728. Banas, Jimmy. "Computer-Controlled Robot Arm." *Radio-Electronics* 56 (May 1985), 49-53.

Provides step-by-step directions for modifying Armatron so that it can be computer controlled. Includes diagrams, photographs, and two sample computer programs. Cover of this issue shows computerized Armatron playing chess, in color.

729. Barbarello, Jim. "Computer-Controlled Robot." *Radio-Electronics* 58 (May 1987), 144-147.

Describes how to build an interface for control of a Radio Shack Mobile Armatron by a Commodore 64 microcomputer. Illustrated with photograph and schematics.

730. Berk, A. A. *Practical Robotics and Interfacing for the Spectrum*. New York: Granada, 1984. 147 p.

Provides introduction to robotics and the Spectrum microcomputer's electronics, and discusses a simple control example, interfaces, robotic peripherals, robot arms, binary, hexadecimal and ASCII Code, electronic components and soldering, and the Spectrum edge connector. Lists sources of components and devices in England.

731. Blankenship, John. *Apple II/IIe Robotic Arm Projects*. Englewood Cliffs, NJ: Prentice-Hall, Inc., 1985. 149 p.

Provides a computer program for a simulated X-Y robotic arm (videobot), instructions for building and computer programs for operating a mechanical X-Y robotic arm, a jointed robotic arm, a robotic arm for a hand puppet, a human-sized robotic arm, and a vision system, directions on how to computerize Armatron, and information on stepper, hobby servo, and DC motors. Includes photographs and drawings.

732. Blankenship, John. "TIMEL: A Homebuilt Robot. Part 1." *Robotics Age* 3 (July/August 1981). Reprinted in *Robotics Age: In the Beginning*, pp. 196-199. Edited by Carl Helmers. Rochelle Park, NJ: Hayden Book Company, Inc., 1983. 241 p.

Describes how to build a robot named TIMEL, using common and inexpensive materials. Includes photographs and drawings.

733. Blankenship, John. "TIMEL: A Homebuilt Robot. Part 2." *Robotics Age* 3 (November/December 1981). Reprinted in *Robotics Age: In the Beginning*, pp. 200-208. Edited by Carl

Helmers. Rochelle Park, NJ: Hayden Book Company, Inc., 1983. 241 p.

Describes TIMEL's internal electronics and its radio link to an external computer. Includes photographs, drawings, and schematics.

734. Bonnett, Kendra R. and Oldfield, Gene. *The Everyone Can Build a Robot Book*. New York: Simon & Schuster, Inc., 1984. 78 p.

Covers history of robots, what a robot is, examples of young people who have built robots, and basic information on electronics. Provides instructions for building a light-seeking, computer-controlled robot called The Moth. Illustrated with photographs and drawings.

735. Boyet, Howard. *HERO 1: Advanced Programming Experiments*. Revised and edited by Ron Johnson. Benton Harbor, MI: Heath Company, 1984. Variable paging.

Designed as a course. Covers Hero's monitor subroutines, machine language control of Hero's I/Os, using Hero's other I/Os, Hero's interpreter, light, sound, and speech applications, motion detection and ultrasonic ranging, Hero's teaching pendant, applications of an industrial or manufacturing nature, Hero reference material, and motor move commands.

736. Brown, Karl. "Build BERT, the Basic Educational Robot Trainer, Part 1." *Byte* 12 (April 1987), 113-118, 120-122.

Provides directions for building a talking, turtle-type, mobile robot, BERT. Includes color photographs and schematics.

737. Brown, Karl. "Build BERT, the Basic Educational Robot Trainer, Part 2." *Byte* 12 (May 1987), 113-119.

Describes how to troubleshoot and program BERT. Includes color photographs.

738. Burkett, Barbara. "Can a Teenager Find Happiness Building Robots?" *Electronic Education* 4 (October 1984), 23.

Describes the work of 13-year-old eighth grader, Scott Dunbar of Tallahassee, Florida, who built a computer-controlled robotic arm.

739. Carroll, Tom. "Build Your Own Robot." *Boy's Life* 77 (February 1987): 18-21.
Provides directions for building GISMO2BL, a mobile robot driven by Milton Bradley's Big Trak motors. Includes photographs, drawings, parts list, and suppliers. See number 811 for examples of the finished product. Cover of this issue shows a color photograph of the robot.

740. Carroll, Tom. "The Smartest Homebuilt Robots; America's Hottest New Robots Are Being Constructed in the Workshops of Amateurs Who Have a Vision for the Future." *Popular Mechanics* 161 (April 1984), 88-90.
Reports on a number of homebuilt, personal robots. Includes color photographs and drawing. Box insert on page 90, "Japan Loves Robot House," describes modular house with computer control system and voice-activated robot, designed by students and faculty at the Illinois Institute of Technology in Chicago.

741. Ciarcia, Steve. "A Computer-Controlled Tank." *Byte* 6 (February 1981), 44-48, 50, 52, 54-55, 58, 60, 62, 64, 66.
Explains how to use a microcomputer to control Big Trak. Illustrated with photographs, drawings, and schematics.

742. Colapinto, John. "The Terminal Man." *Saturday Night* 100 (October 1985), 32-33.
Reports on Roland Knight of Georgetown, Ontario, Canada, and his robot arm. Illustrated with photograph.

743. Connell, Jonathan. "The Omni Photovore: How to Build a Robot That Thinks Like a Roach." *Omni* 11 (October 1988), 201-206, 210, 212.
Describes insect robots built from mobile toy cars at the Artificial Intelligence Laboratory, Massachusetts Institute of Technology, Cambridge, and provides instructions for building

a modified version with suggestions for games to play. Illustrated with photographs, drawings, and schematics.

744. Crypton, Dr. "Machine Wars MIT Style." *Science Digest* 93 (April 1985), 62-67, 81.
Reports on and describes "The Harvest," a four-hour-long, remote-controlled battle of robotic Ping-Pong-ball reapers (harvesters) held annually at Massachusetts Institute of Technology, Cambridge. Includes photographs.

745. Cummings, Richard. *Make Your Own Robots*. New York: David McKay Company, Inc., 1981. 120 p.
Provides directions for making a variety of robots and robot-related objects, including a wind-powered twirling robot, toy robot in a toy car, scooter-like robot, electric rover robot, android mask, writing automaton, and robot costume. Includes drawings and diagrams.

746. Cunningham, John E., and Horn, Delton T. *Handbook of Remote Control & Automation Techniques*. 2nd ed. Blue Ridge Summit, PA: TAB Books, Inc., 1984. 341 p.
Covers basic principles, types of sensing devices, mechanical devices, hydraulic systems, electric motors, various control devices, computer-controlled and radio-controlled systems, and applications to robots.

747. DaCosta, Frank. *How to Build Your Own Working Robot Pet*. Blue Ridge Summit, PA: TAB Books, Inc., 1979. 238 p.
Presents directions for building a wheeled "dog" robot that resembles a small tractor. Includes photographs, drawings, schematics, and computer programs.

748. Ellis, John. "Robot Eyes." *Radio-Electronics* 57 (March 1986), 32, 38.
Describes how to give a robot "eyes" to avoid obstacles.

749. Gargaliano, Tim, and Fons, Kathryn. "Teach Your Robot to Speak." *Robotics Age* 3 (November/December 1981). Reprinted in *Robotics Age: In the Beginning*, pp. 164-170.

Edited by Carl Helmers. Rochelle Park, NJ: Hayden Book Company, Inc., 1983. 241 p.

Covers the basics of synthesizing speech, describes inexpensive hardware that can give a robot unlimited vocabulary, and explains the programming techniques needed to teach a robot how to speak. Illustrated with photographs, tables, and schematics.

750. Gleason, B. J. "Using LOGO for Robot Simulation." *Robotics Age* 7 (September 1985), 27-30.

Describes how to use Atari Logo to simulate a turtle robot, to "build" and test performance of the robot before actual construction, and includes program listings.

751. Gupton, James A., Jr. "Unicorn-1 Robot." *Radio-Electronics* 51 (August 1980), 37-41, 76.

First of a series of articles telling how to build the Unicorn-1, a personal robot. Includes photographs and drawings. Cover of this issue shows a color photograph of the robot.

752. Gupton, James A., Jr. "Unicorn-1 Robot." *Radio-Electronics* 51 (September 1980), 55-58.

Part 2. Shows how to construct the arm and several types of hands for the Unicorn-1 Robot. Includes photographs and drawings.

753. Gupton, James A., Jr. "Unicorn-1 Robot; Assembling the Legs." *Radio-Electronics* 51 (October 1980), 92-95, 126.

Part 3. Describes how to build the mobility base which allows the Unicorn-1 Robot to move around. Includes photographs and drawings.

754. Gupton, James A., Jr. "Unicorn-1 Robot; Assembling the Body." *Radio-Electronics* 51 (November 1980), 61-64.

Part 4. Describes how to construct the body frame and rotation mechanism of the Unicorn-1 Robot. Includes photographs and drawings.

755. Gupton, James A., Jr. "Unicorn-1." *Radio-Electronics* 51 (December 1980), 55-59.

Part 5. Completes description of how to construct the body of the Unicorn-1 Robot, and tells how to control the robot. Includes photographs and drawings.

756. Gupton, James A., Jr. "Unicorn-1 Robot." *Radio-Electronics* 52 (January 1981), 51-55.

Part 6. Describes how to add electronic embellishments - sets of flashing lights, and a wrist end effector to the Unicorn-1 Robot. Includes photographs and drawings.

757. Gupton, James A., Jr. "Unicorn-1 Robot; Adding Remote Control." *Radio-Electronics* 52 (February 1981), 59-63, 97.

Part 7. Tells how to build a relay board for use with radio control of the Unicorn-1 Robot, and describes how to make a robotic arm. Includes photographs and drawings.

758. Gupton, James A., Jr. "Unicorn-1 Robot." *Radio-Electronics* 52 (March 1981), 63-66, 82.

Part 8. Continues description of how to build a remote control system for the Unicorn-1 Robot, describing control scheme, FM transmitter, and relay-driven board.

759. Gupton, James A., Jr. "Unicorn-1 Robot." *Radio-Electronics* 52 (April 1981), 65-69, 88.

Part 9. Concludes description of remote control system of Unicorn-1 Robot, describing tone-encoder, tone-decoder, and latch boards. Includes photographs and drawings.

760. Gupton, James A., Jr. "Computer Control for the Unicorn-1 Robot." *Radio-Electronics* 52 (May 81), 53-55.

Part 10. Describes methods of computer control, interfacing, and programming for the Unicorn-1 Robot. Includes photographs and drawings.

761. Gupton, James A., Jr. "Unicorn-1 Robot; How to Add Sensors." *Radio-Electronics* 52 (June 1981), 57-60.

Part 11. Concludes description of how to build the Unicorn-1 Robot, describing how to add contact and proximity sensors. Includes photographs and drawings.

762. Heiserman, David L. *Build Your Own Working Robot*. Blue Ridge Summit, PA: TAB Books, Inc., 1976. 234 p.
Provides instructions on how to build Buster I, II, and III robot systems. Includes photographs, drawings, and schematics.

763. Heiserman, David L. *Build Your Own Working Robot; the Second Generation*. Blue Ridge Summit, PA: TAB Books, Inc., 1987. 130 p.
Provides instructions on how to build an updated version of the robot Buster. Includes drawings and schematics.

764. Heiserman, David L. *How to Build Your Own Self-Programming Robot*. Blue Ridge Summit, PA: TAB Books, Inc., 1979. 237 p.
Provides instructions on how to build Rodney Robot, using the 8085 microprocessor. Includes computer programs and flowcharts. Instructions for adding a cassette tape interface and expansion of the system are given. Illustrated with photographs and schematics.

765. Heiserman, David L. *How to Design and Build Your Own Custom Robot*. Blue Ridge Summit, PA: TAB Books, Inc., 1981. 462 p.
Covers technical information needed to build a hobby robot, including electrical and mechanical subsystems, power supplies, motors, speed and internal controls, sensing mechanisms, and microprocessors, and provides examples of a number of robots. Includes computer programs. Illustrated with drawings and schematics.

766. Heiserman, David L. *Projects in Machine Intelligence for Your Home Computer*. Blue Ridge Summit, PA: TAB Books, Inc., 1982. 337 p.

Lists and explains computer programs that demonstrate self-programming machine intelligence. A sequel to *Robot Intelligence...with Experiments*.

767. Heiserman, David. *Robot Intelligence...with Experiments*. Blue Ridge Summit, PA: TAB Books, Inc., 1981. 322 p.

Examines machine intelligence and shows how to simulate robot "thinking" through three progressively higher classes of computer programs in project format. Includes games to test a robot's learning ability and how it adapts to the environment. Lists programs. Provides diagrams and flowcharts.

768. Loofbourrow, Tod. *How to Build a Computer-Controlled Robot*. Rochelle Park, NJ: Hayden Book Company, Inc., 1978. 132 p.

Provides instruction for building "Mike," a microprocessor-controlled robot with voice recognition, impact sensors, and ultrasonic detection. Includes photographs, drawings, schematics, and computer programs.

769. Mandell, Alan. "Using the Microcomputer to Simulate Machine Intelligence." *Journal of Computers in Mathematics and Science Teaching* 4 (Winter 1984), 56, 86.

Discusses the book *Projects in Machine Intelligence for Your Home Computer*, by David L. Heiserman, and provides a sample computer program written by Mandell for the TRS-80, Model III. Program involves a robot that destroys aliens, illustrating the "first law of robotics," that a robot must be programmed with the command that none of its actions will be harmful to humans.

770. McAllister, Don. "Multiple Sensors for a Low-Cost Robot." *Robotics Age* 2 (Spring 1980). Reprinted in *Robotics Age: In the Beginning*, pp. 153-158. Edited by Carl Helmers. Rochelle Park, NJ: Hayden Book Company, Inc., 1983. 241 p.

Describes a method of constructing a multi-channel, analog-to-digital converter that can be used with the Radio Shack TRS-80 computer, allowing programs written in Level

II BASIC to acquire input from several sensors. Illustrated with photographs, charts, and schematics.

771. McComb, Gordon. *The Robot Builder's Bonanza; 99 Inexpensive Robotic Projects*. Blue Ridge Summit, PA: TAB Books, 1987. 326 p.

Tells how to build the parts that make up a hobby robot. Covers robot basics, body and frame, power and locomotion, appendages, eyes, ears, and mouth, navigation, and electronic control, and presents instructions on projects that can be built. Appendices list supply sources and further reading, provide a drill bit and bolt chart, and give information on interfacing logic families and integrated circuitry. Illustrated with photographs, drawings, and schematics.

772. Nagle, Timothy, and Powers, V. Michael. "A Robot Arm without a Budget." *Robotics Age* 3 (January/February 1981). Reprinted in *Robotics Age: In the Beginning*, pp. 180-185. Edited by Carl Helmers. Rochelle Park, NJ: Hayden Book Company, Inc., 1983. 241 p.

Describes part of a project to build a low-cost robot arm or manipulator. Provides photographs and drawings.

773. Nansel, Bob. "The LIMBO Project." *Micro Cornucopia* no. 47 (May/June 1989), 8-12, 14.

Part 1 of 4. Provides background information on hobby robots, on building LIMBO, a maze-running robot, and on Seattle Robotics Society Maze Contest. Illustrated with photographs and color drawing.

774. Nansel, Bob. "LIMBO Motors On." *Micro Cornucopia* no. 48 (July/August 1989), 28-33.

Part 2 of series on how to build LIMBO, a maze-running robot. Illustrated with photographs and drawings.

775. Nansel, Bob. "LIMBO Part Three." *Micro Cornucopia* no. 50 (November/December 1989), 22-26, 28-31.

Part 3 of series on how to build LIMBO, a maze-running robot. Illustrated with photographs and drawings.

776. Nansel, Bob. "LIMBO, Part Four." *Micro Cornucopia* no. 51 (January/February 1990), 30-32, 34, 36, 38-41.
Part 4 of series on how to build LIMBO, a maze-running robot. Illustrated with drawings and schematics.

777. Netsel, Tom. "Closet Computers." *Compute!* 10 (October 1988), 14-16, 18, 20.
Includes description of ways old personal computers are used to control robots built by Gordon McComb of Los Angeles, California. Illustrated with drawings.

778. Otten, David. "Building MITEE Mouse III." *Circuit Cellar Ink* (June/July 1990), 32-39.
Describes how to build the Massachusetts Institute of Technology Electrical Engineering Mouse. Illustrated with drawings and schematics.

779. Parrett, Tom. "The Rise of the Robot." *Science Digest* 91 (April 1983), 68-75, 107.
Reports on hobby roboticists (or "robotics hobbyists"), and the robots they built or worked with. Illustrated with color photographs. Cover of this issue features color drawing of a robot.

780. Pattis, Richard E. *Karel the Robot; A Gentle Introduction to the Art of Programming*. New York: John Wiley & Sons, 1981. 106 p.
Uses a computer-simulated robot as an introduction to learning the program language PASCAL. An interesting project on its own.

781. Potter, Tony. *How to Make Computer-Controlled Robots*. Illustrated by Jeremy Gower and Simon Roulstone. London: Usbourne Publishing, Ltd., 1984. 48 p.
An elementary-level how-to-do-it book for young people, useful for older people too. Provides step-by-step instructions, fully illustrated with color drawings. Includes templates for transfer to balsa wood, and a computer-control program. The finished robot is an arm type with gripper, mounted on a

wheeled base to make it a mobile vehicle. The robot was designed by Tony Potter and Chris Oxlade; the robot computer program was written by Chris Oxlade.

782. Robillard, Mark J. "Building a Robot from the Bottom Up." *Radio-Electronics* 56 (September 1985), 90, 92-93.
Describes how to build a programmable motion platform as a mobility base for a robot. Illustrated with drawings.

783. Robillard, Mark J. "Building Your Own Robot." *Radio-Electronics* 57 (March 1986), 47-50.
Gives general directions for building a robot. Includes information on helpful books, robot kits, and parts suppliers. Illustrated with photographs and diagrams.

784. Robillard, Mark J. "Experimental Robot Vision." *Radio-Electronics* 57 (December 1986), 81-83.
Describes how to connect a photocell-based vision sensor to a microcomputer for use with a robot.

785. Robillard, Mark J. "Getting Started in Robotics." *Radio-Electronics* 56 (July 1985), 66, 70.
Discusses personal and industrial robots and the industrial future. Includes photograph of Hero 1. First in a series of columns on robots.

786. Robillard, Mark J. "Inexpensive Robotics Arms." *Radio-Electronics* 57 (July 1986), 80-81.
Describes how to build a simple robot arm using Milton Bradley's Robotix building kit. Includes drawing.

787. Robillard, Mark J. "Mystery Computer Revealed." *Radio-Electronics* 57 (April 1986), 81, 104.
Discusses computer control of robots and the use of Radio Shack's Model 100 microcomputer for this control.

788. Robillard, Mark J. "Position Sensing." *Radio-Electronics* 57 (September 1986), 74-76.

Discusses digital position sensing, a reflectance decoder, and analog sensing for robots.

789. Robillard, Mark J. "Robot Brains." *Radio-Electronics* 57 (February 1986), 94-95.
Discusses computer control of home-built robots, and a possible design for a computer to control robots.

790. Robillard, Mark J. "Robots and Voice Recognition." *Radio-Electronics* 57 (January 1986), 84, 86.
Discusses voice recognition and describes one system, EARS (Speech Systems, 38W 255 Deerpath Road, Batavia, IL 60510) that can be used with a home-built robot.

791. Robillard, Mark J. "Setting up a Robotics Design Lab." *Radio-Electronics* 56 (August 1985), 80-81.
Discusses lab layout, floor layout, type of robot to build, and control systems. Illustrated with drawings.

792. Robillard, Mark J. "Simple Solid-State Vision." *Radio-Electronics* 57 (November 1986), 74-75.
Describes the design of a robot vision system, a photocell-based vision system that can be built for a robot, and pattern recognition.

793. Robillard, Mark J. "Tactile Sensing." *Radio-Electronics* 57 (August 1986), 71-72.
Discusses touch, optical, and force sensors for robots.

794. Robillard, Mark J. "Ultrasonic Rangers and Stepper Motors." *Radio-Electronics* 56 (December 1985), 88, 113.
Discusses stepper motors, and robot mapping using a stepper motor and an ultrasonic ranger. Illustrated with schematic.

795. Robillard, Mark J. "Ultrasonic Vision." *Radio-Electronics* 56 (November 1985), 89-90, 108.

Discusses machine-vision systems, ultrasonics theory, and circuits for use as a feedback mechanism for a robot, and Polaroid's ultrasonics.

796. Ruby, Daniel. "Bump-and-Bounce Robot Kits." *Popular Science* 223 (October 1983), 160.
Describes Robot Shack's robot kits, X-1 and Droid-Bug. Illustrated with photographs.

797. Sarns, Steven E. "Build This R-E Robot. Part 1." *Radio-Electronics* 57 (December 1986), 54-56, 94.
Begins a series of 13 articles on how to build and use a mobile, "Radio-Electronics" (R-E) robot. Cover of this issue shows a drawing of the robot, with controlling circuitry superimposed on a young man's head.

798. Sarns, Steven E. "R-E Robot. Part 2." *Radio-Electronics* 58 (January 1987), 42-44, 69.
Describes the Robotic Personal Computer used in the R-E Robot. Illustrated with schematics and diagrams.

799. Sarns, Steven E. "R-E Robot. Part 3." *Radio-Electronics* 58 (February 1987), 48-50.
Discusses design and construction of the R-E Robot drive system. Illustrated with drawings.

800. Sarns, Steven E. "R-E Robot. Part 4." *Radio-Electronics* 58 (March 1987), 52-56.
Shows how to build the R-E Robot's body and describes an updated drive system. Illustrated with photographs and drawings.

801. Sarns, Steven E. "R-E Robot. Part 5." *Radio-Electronics* 58 (April 1987), 39-44, 102, 108.
Provides hardware details on the Robotic Personal Computer for the R-E Robot. Illustrated with diagrams and schematics.

802. Sarns, Steven E. "R-E Robot. Part 6." *Radio-Electronics* 58 (May 1987), 62-66.

Provides an in-depth look at the R-E Robot's control electronics. Illustrated with photograph and schematics.

803. Sarns, Steven E. "R-E Robot. Part 7." *Radio-Electronics* 58 (June 1987), 58-60, 90.

Looks at the R-E Robot's control electronics. Illustrated with diagrams and photograph.

804. Sarns, Steven E. "R-E Robot. Part 8." *Radio-Electronics* 58 (July 1987), 44-46, 78.

Describes how to build the R-E Robot's control board. Illustrated with photographs and schematics.

805. Sarns, Steven E. "R-E Robot. Part 9." *Radio-Electronics* 58 (August 1987), 57-60, 80.

Describes the R-E Robot's command language, RCL (Robotic Control Language). Provides a computer program and diagram for an experimental trip through an office. Illustrated with photograph and drawing.

806. Sarns, Steven E. "R-E Robot. Part 10." *Radio-Electronics* 58 (September 1987), 56, 59-60, 63, 86.

Explains applications of the R-E Robot that demonstrate how to develop its computer programs, read the A/D converter and the switch status, and control motor speed and direction. Illustrated with photograph and diagram.

807. Sarns, Steven E. "R-E Robot. Part 11." *Radio-Electronics* 58 (October 1987), 56-59, 74.

Tells how to add a robotic arm to the R-E Robot and how to control it. Illustrated with photographs and schematics.

808. Sarns, Steven E. "R-E Robot. Part 12." *Radio-Electronics* 58 (December 1987), 67-68, 84.

Discusses the addition of an electronic eye to the R-E Robot. Illustrated with schematic, diagrams, and photograph.

809. Sarns, Steven E. "R-E Robot. Part 13." *Radio-Electronics* 59 (January 1988), 59-62.
Last in the series on the R-E Robot. Describes a head unit that serves as an interface for the navigational and sensory circuits, including the eye, of the R-E Robot. Illustrated with diagrams and schematics.

810. Shircliff, David R. *Build a Remote-Controlled Robot for Under $300*. Blue Ridge Summit, PA: TAB Books, Inc., 1986. 119 p.
Provides an introduction to robots and tells how to build Questor, a motorized, wheeled robot. Illustrated with photographs, drawings, and schematics.

811. Stuckey, Scott. "We Built GISMO2BL." *Boy's Life* 79 (February 1989), 38-39.
Reports on robots built using directions given in the February 1987 *Boy's Life* article, "Build Your Own Robot." Includes color photographs of some of the robots built. See number 739 for original instructions.

812. Tavares, J. A. "Light-Seeking Robot." *Radio-Electronics* 57 (November 1986), 38-39.
Describes how to build a circuit for demonstrating light seeking by a robot.

813. Thyng, Michael. "About Robots: An Account of Costs." *C.A.L.L. Apple* 7 (May 1984), 10-12.
Discusses the costs involved in building a robot, the design used by the author, and where problems arose. Includes partial list of materials used.

814. Weinstein, Martin Bradley. *Android Design; Practical Approaches for Robot Builders*. Hasbrouck Heights, NJ: Hayden Book Company, 1981. 248 p.
Discusses and describes the building of hobby robots, including philosophical considerations. Covers obstacles in the human environment, the problem of stairways, designing the main mechanical drive, chassis, main motor drive, battery,

motors, heat removal, collision avoidance, fingers, vision, mapping, voice, and brains. Provides addresses of parts suppliers throughout. Illustrated with photographs, drawings, and schematics.

815. Weinstein, Martin Bradley. "Design Your Own Android. Part 1." *Radio-Electronics* 51 (January 1980), 37-40.

Describes what an android is, goals of an android, obstacles in the human environment, physical aspects of an android, size and weight, arm design problems, and other considerations. Lists sources of parts and information. Illustrated with drawings and photograph. Cover of this issue shows color photograph of R2D2 robot, with and without outer casing, manufactured by Kenner Products.

816. Weinstein, Martin Bradley. "Design Your Own Android. Part 2." *Radio-Electronics* 51 (February 1980), 53-56.

Defines the problem of android vision and discusses depth perception, considers track drive design, stairway geometry, tight corners, leaning and trunk rotation, and other problems of android design. Illustrated with drawings and photograph.

817. Wingerson, Lois. "Robot Junkies." *Science '84* (June 1984), 56, 58-63.

Describes some of the robots built by members of the Homebrew Robotics Club of Palo Alto, California, and activities of these members. Illustrated with photographs.

Toy Robots

818. "It Walks, Talks and Falls Asleep, and So Animal Lovers Must Ask - Is It Live or Is It Robopet?" *People Weekly* 28 (December 7, 1987), 97.

Describes two toy robot dogs invented by Bill Gillard and sold by his company Phonetica One, Colorado City, Colorado. Includes photograph.

819. "Robots from TV." *Fortune* 115 (February 2, 1987), 8-9.
Reports on Axlon Incorporated's programmable robot toys that obey tones embedded in television cartoon broadcasts or in videotapes. Includes color photograph.

820. Arrington, Carl, and Velasco, Irma. "Deck the Halls with Squads of Robots: Hasbro Takes on Tonka in the Toy Wars of 1984." *People Weekly* 22 (December 3, 1984), 154-155, 157.
Reports on the transformable toy robot market, particularly GoBots and Transformers made by Hasbro and Tonka.

821. Bailly, Christian, and Bailly, Sharon. *Automata: The Golden Age, 1848-1914.* London: Sotheby's Publications, 1987. 360 p.
Covers automata industry of Paris, France, the master automaton makers, inside the automaton, and restoration, and includes makers' catalogs in facsimile. Illustrated with color and black-and-white photographs.

822. Brophy, Rick. "Toy Robots." *Antiques & Collecting Hobbies* 93 (November 1988), 64-67.
Discusses history of toy robot collecting and how to collect them. Illustrated with photographs.

823. Creanza, Rose. "The Once and Future Robots." *Working Woman* 7 (July 1982), 43.
Reports on toy robots as collectibles and on uses and capabilities of robots. Illustrated with color photographs of a collection of toy robots.

824. Dodd, Lynn Strongin. "Robots on the Move." *Financial World* 154 (August 7, 1985), 47.
Reports on transformable robot toys. Illustrated with drawing.

825. Gardner, Howard. "Toys with a Mind of Their Own." *Psychology Today* 13 (November 1979), 93-95, 97-98, 101.

Describes a number of electronic toys, and includes description and color photograph of 2XL, talking toy robot from MEGO that asks questions and gives answers using an 8-track cassette tape.

826. Griffith, F. H. "Mechanical Banks & Toys: Robots." *Hobbies* 88 (February 1984), 41-42.
Discusses collecting toy robots. Includes photographs.

827. Griffith, F. H. "Robots." *Hobbies* 85 (March 1980), 52-53.
Describes a number of battery-operated toy robots. Includes photographs.

828. Griffith, F. H. "Robots." *Hobbies* 85 (April 1980), 54-55.
Describes a number of battery-operated toy robots. Includes photographs.

829. Griffith, F. H. "Robots." *Hobbies* 85 (May 1980), 54-55.
Describes a number of battery-operated toy robots. Includes photographs.

830. Griffith, F. H. "Robots." *Hobbies* 85 (June 1980), 54-55.
Describes a number of toy robots. Includes photographs.

831. Griffith, F. H. "Robots." *Hobbies* 85 (September 1980), 54-55.
Describes toy robots sold in Sears, Roebuck and Company's 1960 Christmas Catalog. Includes photographs.

832. Griffith, F. H. "Robots." *Hobbies* 85 (October 1980), 54-55.
Describes toy robots sold in Sears, Roebuck and Company's Christmas catalogs. Includes photographs.

833. Griffith, F. H. "Robots." *Hobbies* 85 (November 1980), 57.
Describes toy robots sold in Sears, Roebuck and Company's 1966 Christmas Catalog. Includes photographs of advertisements.

834. Griffith, F. H. "Robots." *Hobbies* 85 (December 1980), 58-59.

Last in series on toy robots sold in Sears, Roebuck and Company's Christmas catalogs. Includes illustrations of advertisements.

835. Griffith, F. H. "Robots." *Hobbies* 87 (January 1983), 58-59.
Discusses collecting toy robots. Includes photographs.

836. Griffith, F. H. "Robots." *Hobbies* 87 (February 1983), 55-57.
Discusses collecting toy robots. Includes photographs.

837. Griffith, F. H. "Robots." *Hobbies* 88 (January 1984), 52-53.
Discusses collecting toy robots. Includes photographs.

838. Griffith, F. H. "Robots." *Hobbies* 89 (May 1984), 44-45.
Discusses collecting toy robots. Mentions that toy astronauts or spacemen are classified or listed with robots. Also mentions the group of robots known as planet robots. Includes photographs.

839. Griffith, F. H. "Robots." *Hobbies* 89 (June 1984), 42-43.
Discusses collecting toy robots. Includes photographs.

840. Kitahara, Teruhisa. *Robots: Tin Toy Dreams*. San Francisco, CA: Chronicle Books, 1985. 111 p.
Provides color photographs and descriptions of 229 toy tin robots manufactured in the 1950's and 1960's, from the author's collection of over 10,000 tinplate toys in Japan.

841. Koepp, Stephen, and Buder, Robert. "Buddy System; Woz and King Pong Team Up." *Time* 127 (May 5, 1986), 50.
Reports on the teaming of Nolan Bushnell and Stephen Wozniak to produce robot toys through Axlon, Incorporated. Includes color photograph.

842. McKinney, Kevin. "Toybot." *Omni* 9 (January 1987), 52-57.
A color photograph gallery of toy robots of the 1950's and 1960's in action, dioramas arranged by artist Douglass Grimmett and photographer Chuck Carlton.

843. Moran, Brian. *Battery Toys: The Modern Automata.* Exton, PA: Schiffer Publishing, Ltd., 1984. 166 p.

Covers collecting battery-operated toys, care and repair, and various types of battery toys, including robots, and provides a collector's desirability chart. Illustrated with color and black-and-white photographs.

844. Reed, J. D., and White, Sara. "Hot Toys with a Special Twist; Tricky Robots Capture Little Kids and Big Bucks." *Time* 124 (October 1, 1984), 73.

Reports on the transformable toy robot market. Includes color photographs.

845. Rogers, Michael and Atkins, Sid. "Toy Soldiers Go High-Tech." *Newsweek* 107 (May 5, 1986), 54, 57.

Reports on Tech Force and other toy, television-controlled and remote-controlled robots built by Axlon, Incorporated. Includes color and black-and-white photographs.

846. Ross, Dave. *Making Robots.* New York: Franklin Watts, 1980. 32 p.

Provides directions for making toys resembling robots, using materials around the house. Illustrated with color drawings.

SECTION 5. NONEDUCATIONAL ROBOT USES

Robot applications of all types other than educational uses are covered in this section. The largest grouping, Industrial Robots, contains materials describing robots whose chief purposes are related to industry. Reports suggest that perhaps half of all industrial robots in use today worldwide are being used in the automobile industry, especially in spot welding, arc welding, painting, and assembly. However a growing number of other industries are making use of robots and they can be found assembling products as vastly different as wristwatches and locomotives. School vocational programs which focus on industrial careers should include study of industrial robotics.

A wide variety of uses ranging from robot food handlers in restaurants to robot duck decoys can be found described in materials under the Other Robot Uses grouping. Some of the materials deal with industrial robots, such as those covering the aircraft industry, but are placed in this section to keep them conveniently located with materials on similar topics. This grouping includes specific industrial, service, and miscellaneous robot uses, such as agricultural, entertainment, health care and medical, military, nuclear and other hazardous environments, sports, and underwater. Service robots are considered to be distinct from industrial robots and have been defined by the National Service Robot Association to be "Machines that sense, think and act to benefit or extend human capabilities and to increase human productivity." Educational robots are considered to be service robots. Robots for use with older people, in health care, and as an aid to the handicapped are clearly service robots.

Materials listed in this section illustrate the great versatility of robots and will provide students with ideas for ways robots may be used in the future.

INDUSTRIAL ROBOTS

General Information

847. "A Personnel Manager's Guide to Robots." *The Futurist* 15 (February 1981), 35.

Summarizes information on industrial robots from *Robotics in Practice: Management and Applications of Industrial Robots* by Joseph F. Engelberger.

848. "Completing an Automation Laboratory." *Science News* 131 (January 10, 1987), 25.

Reports on the Automated Manufacturing Research Facility at the National Bureau of Standards, Gaithersburg, Maryland. The facility is used as a research laboratory for studying and testing methods of linking computer-controlled machines, which include industrial robots.

849. "Industrial Robots Come of Age." *Dun's Business Month* 123 (February 1984), Special Report, pp. V, X, Z.

Provides an overview on industrial robots, background, choosing and installing a robot, and how robots are controlled.

850. "Ironing Out the Bugs." *USA Today* 110 (February 1982), 12.

Reports on research being done to make industrial robots more effective and efficient. Includes drawing.

851. "Japan and US Lead World's Robot League." *New Scientist* 97 (February 17, 1983), 469.

Surveys the number of robots used in 1982 in countries around the world, analyzes applications in Great Britain, and reports on robot manufacturers there.

852. "Robotics in the UK." *The Industrial Robot* (March 1981), 32-38.

Reports on industrial robot use in Great Britain.

853. "Robots: Japan's Answer to Guest Workers." *Science News* 122 (November 6, 1982), 298-299.
Reports on production and use of industrial robots in the United States and Japan. Includes photograph.

854. "Robots: Why U. S. Firms Buy Japanese." *Newsweek* 99 (March 15, 1982), 61.
Reports on United States and Japanese industrial robotics markets.

855. "Robots Are Coming to Industry's Service." *The Economist* 280 (August 29, 1981), 71-75.
Covers what a robot is, types of industrial robots, types of robot arms, how robots work, applications, and robot manufacturers. Includes photograph. Cover of this issue shows a robot arm in action.

856. "Robots Bump into a Glutted Market." *Business Week* no. 2784 (April 4, 1983), 45.
Reports on a glut of industrial robots in 1982 and 1983. Includes drawing.

857. "Robots Join the Labor Force." *Business Week* no. 2640 (June 9, 1980), 63-65, 68, 73.
Reports on production and use of industrial robots. Includes photographs.

858. "Russian Robots Run to Catch Up." *Business Week* no. 2701 (August 17, 1981), 120.
Reports on production and use of industrial robots in Soviet Union. Includes photograph.

859. "The Factory of the Future." *USA Today* 113 (April 1985), 13.
Reports on predictions about the factory of the future, including the role of robots. Illustrated with photograph of robot at Purdue University, West Lafayette, Indiana.

860. "The Push for Dominance in Robotics Gains Momentum." *Business Week* no. 2718 (December 14, 1981), 108-109.
Reports on production of industrial robots in 1980 and 1981. Includes photograph.

861. "The Revolution That Wasn't." *World Press Review* 34 (August 1987), 48.
Discusses the problems of implementing industrial robots. Illustrated with drawing.

862. "The Risks of Robotization." *The Futurist* 21 (May/ June 1987), 56.
Summarizes findings from a study on industrial robot use worldwide by the International Labour Organisation. Illustrated with photograph.

863. "The Road to Robotics." *Management Today* (January 1984), 68-71.
Describes use of industrial robots in England and why more are not used. Illustrated with photographs and drawing.

864. "The Robot Revolution That May Be a Mirage." *Business Week* no. 2764 (November 8, 1982), 52-53.
Reports on use of industrial robots in Soviet Union, and the 1982 Industrial Robots Trade Show there. Includes photograph.

865. "The Robots Are Coming." *Barron's* 63 (April 11, 1968), 8-9, 37.
Discusses the industrial robotics industry and some of the companies involved in manufacturing industrial robots. Illustrated with photographs.

866. "The State of Robotics in Italy at the Start of the 80's." *The Industrial Robot* (September 1981), 176-177.
Reports on the industrial robot's use in Italy.

867. "The West's Risky Runup in Robot Sales." *Business Week* no. 2725 (February 8, 1982), 44-45.
Reports on sales of industrial robots to Soviet Union by United States and Japanese companies and applications there. Includes photograph.

868. "What Does the Future Hold?" *USA Today* 112 (June 1984), 9-10.
Reports on the potential of robots in industry. Illustrated with photographs of The Market-Robot, a custom-designed robot which can broadcast messages and carry on conversations with passers-by.

869. "What Next? Liability Insurance for Robots!" *Production Engineering* 31 (March 1984), 10.
Describes "personal liability" insurance available to cover industrial robots against damage and related problems.

870. "When Robots Take Over People's Jobs." *U. S. News & World Report* 90 (February 16, 1981), 75-77.
Reports on uses of industrial robots and provides a graph on the world's robot population. Illustrated with photographs.

871. Albus, James S. "Robots in the Workplace." *The Futurist* 17 (February 1983), 22-27.
Discusses present and future uses of robots in industry, and some of the problems involved in implementing robot technology. Illustrated with photographs.

872. Austen, Ian, and Fleming, James. "Canada Confronts the Robotics Age." *Maclean's* 95 (November 8, 1982), 44-45.
Reports on robot use in industry in Canada, Japan, United States, and Western Europe. Includes color and black-and-white photographs.

873. Ayres, Robert, and Miller Steve. "Industrial Robots on the Line." *Technology Review* 85 (May/June 1982), 35-36, 38-44, 46.

Discusses robot applications, history, installations, the future factory, union response to robots, and displaced worker training. Includes illustrated chronology of significant developments in robotics technology. Illustrated with photographs and drawings.

874. Ayres, Robert U., and Miller, Steven M. "Robotic Realities: Near-Term Prospects and Problems." *Annals. American Academy of Political and Social Science* 470 (November 1983), 28-55.

Describes industrial robots, discusses applications, lists robot capabilities, benefits, efficiency-flexibility trade-off, potential impacts of labor cost and capacity, inventory, work-in-process and capital costs, the consequences of dramatic improvements, employment effects, and job openings.

875. Ayres, Robert U., and Miller, Steven M., et al. *Robotics and Flexible Manufacturing Technologies; Assessment, Impacts, and Forecast*. Park Ridge, NJ: Noyes Publications, 1985. 443 p.

Assesses, discusses, and makes forecasts for the use of industrial robots and flexible manufacturing technologies, and covers current robotic technology and research and development activities in robotics.

876. Bock, Gordon; Ishikawa, Yukinori; and McCarroll, Thomas. "Limping Along in Robot Land." *Time* 130 (July 13, 1987), 46-47.

Reports on the industrial robot industry and use of industrial robots. Includes color drawing.

877. Button, Rick. "Robotics." In *Proceedings of the Statewide Conference on High Technology, Austin, Texas, October 7-8, 1982*, pp. 24-32. Austin: Texas Advisory Council for Technical-Vocational Education and Texas Education Agency, 1983. 95 pp. ERIC document ED 226 128.

Discusses robot parts, interfacing, sensors, implementation, industry structure, power devices, computers and controls, tooling, applications, and skills needed.

878. Bylinsky, Gene. "Those Smart Young Robots on the Production Line." *Fortune* 100 (December 17, 1979), 90-93, 95-96.

Although now dated in some areas, provides a useful introduction to industrial robots and their applications. Includes color photographs.

879. Callahan, J. Michael. "The State of Industrial Robotics; What Today's Robots Can Do and What the Future Holds." *Byte* 7 (October 1982), 128, 130, 132, 134, 136-140, 142.

Covers the definition of industrial robot, historical background, robot fundamentals, systems and applications, and the future. Includes a chart of specifications for seven industrial robots, and a list of names and addresses of industrial robot vendors. Illustrated with drawings and color photographs.

880. Cardoza, Anne, and Vlk, Suzee J. *Robotics*. Blue Ridge Summit, PA: TAB Books, Inc., 1985. 149 p.

Covers what robotics includes, how and why robots are used in industry, some of the specific robots used, job opportunities, and robotics education and training. Appendixes include lists of robotics degree programs, historical timetable, and manufacturers. A glossary and bibliography is provided. Illustrated with photographs.

881. Chamberlain, Gary. "Robotics: No Humanoids Yet, But...." *Design News* 43 (March 23, 1987), 26-27.

Reports on robots being developed for a number of industrial uses. Illustrated with color and black-and-white photographs.

882. Cohen, Peter A. "What Became of the Projected Robotics Market Boom?" *Computerworld* 19 (May 27, 1985), SR3-SR4.

Discusses reasons why industrial robot market did not grow as fast as expected. Includes graphs.

883. Copulos, Milton. "The Factory of the Future." *Consumers' Research Magazine* 69 (July 1986), 25-28.
Describes the uses of automation, including robots, in factories.

884. Cromie, William J. "Robots: A Growing Maturing Population." *SciQuest* 54 (March 1981), 12-16.
Discusses advantages and applications of industrial robots. Includes color photographs. Cover of this issue shows color photograph of industrial robot.

885. Donlan, Thomas G. "Robot Explosion." *Barron's* 62 (April 5, 1982), 8-9, 20.
Discusses the industrial robot industry and some of the companies that manufacture industrial robots. Illustrated with photographs.

886. Dorf, Richard C. *Robotics and Automated Manufacturing.* Reston, VA: Reston Publishing Company, Inc., 1983. 190 p.
Includes robot history, development, and classification, components and operation of robot systems, types of robotic systems, vision sensors and systems, artificial intelligence, robot applications, economic considerations, worker considerations, and the future of robotics and automated manufacturing. Illustrated with photographs and drawings.

887. Dunne, Maurice J. "Robotics: Past, Present, and Future." Paper presented at the American Vocational Association Convention, St. Louis, Missouri, December 5, 1982. 37 p. ERIC document ED 224 978.
Relates the history of robot use in industry and discusses applications and reasons to use, advances in the field, and programs and curriculums being developed. Illustrated with photographs, charts, and tables.

888. Edelson, Edward. "Robots - the Next Generation." *Popular Science* 226 (February 1985), 42, 44-46.
Discusses problems in designing more advanced industrial robots.

889. Engelberger, Joseph F. *Robotics in Practice; Management and Applications of Industrial Robots*. New York: AMACOM, 1980. 291 p.

Considers robot use in manufacturing, robot anatomy, end effectors, matching robots to the workplace, reliability, maintenance and safety, organizing to support robots, robot economics, sociological impact, and future capabilities. Provides a section on application studies. Includes color and black-and-white photographs, charts, drawings, and a bibliography.

890. Feulner, Terry, and Kleiner, Brian H. "When Robots Are the Answer." *Personnel Journal* 65 (February 1986), 44-47.

Discusses jobs robots can perform better than humans, implementation of a robotic system in industry, and addressing human relations concerns. Illustrated with color photographs.

891. Friedrich, Otto; Redman, Christopher; and Simpson, Janice C. "The Robot Revolution." *Time* 116 (December 8, 1980), 72-78, 83.

Reports on the use of industrial robots in manufacturing, and on projects to improve industrial robots. Includes color photographs and a cartoon.

892. Froehlich, Lee, and Klee, Ken. "Robotics: Technology's Heir Apparent." *Savvy* 4 (March 1983), 44-48.

Discusses the industrial robot market in the United States and Japan. Illustrated with photographs and drawings.

893. Froehlich, Leopold. "Mighty Oaks Take Time." *Datamation* 29 (January 1983), 113-114, 116, 118.

Discusses potential and predictions for robots in industry, Japanese-United States agreements, Japanese and United States government support of the robotics industry, specific uses by United States companies, need for interface standards, and growing student interest in robotics. Illustrated with drawing.

894. Froehlich, Leopold. "Robots to the Rescue?" *Datamation* 27 (January 1981), 84-86, 88, 90, 94, 96.
Reports on the Robots V Conference held October 1980 in Dearborn, Michigan, discusses general productivity factors which lead to use of industrial robots, and explains what an industrial robot is and the classes of industrial robots. Includes color photograph. Cover of this issue shows robot arm in motion.

895. Fu, King-sun. "Robotics and Automation." *Computer* 15 (December 1982), 13-14.
Introduces a group of six articles on industrial robots in this issue. Includes color photograph.

896. Garrison, H. Don. "Robots Are Taking Over--Who Does What." Paper presented at the American Vocational Association Convention, New Orleans, Louisiana, December 1984. 19 p. ERIC document ED 252 676.
Describes industrial robots and applications, and methods for determining feasibility and correct robot type for specific applications.

897. Gengler, Michele J., and Tersine, Richard J. "Robots - Coming to Work in America." *Business* 29 (May/June 1983), 3-12.
Discusses advantages of robots, applications, the new generation of robots, problems to be solved, the automated factory, and the transition in the work force. Includes robot genealogy and predictions for the future.

898. Gregory, Gene. "The Factory of Japan's Future." *Management Today* (April 1984), 66-71.
Reports on robots used in industry, especially in Japan. Illustrated with photographs.

899. Hartley, John. *Flexible Automation in Japan*. Kempston, Bedford, England: IFS (Publications) Ltd., 1984. 264 p.
Covers robots and their applications in industry in Japan. Includes photographs, drawings, and charts.

900. Hartley, John. *Robots at Work; a Practical Guide for Engineers and Managers*. Kempston, Bedford, England & Amsterdam: IFS (Publications) Ltd. & North-Holland Publishing Company, 1983. 191 p.

Covers why robots are used, robot types, robot control, cost-effectiveness, product design for robot manufacture, safety, flexible manufacturing systems, robot applications in spot welding, handling, arc welding, spraying, assembly, sensing, and other jobs, robot trolleys, and where robots are taking us. Illustrated with photographs and drawings.

901. Holzbock, Werner G. *Robotic Technology*. New York: Van Nostrand Reinhold Company, 1986. 494 p.

Covers statics and dynamics, robot configurations, mechanical parts, hydraulic concepts, systems and components, pneumatics, servomotors, feedback sensors, electronic hardware, programming and controlling, robot vision, and safety. Illustrated with drawings and photographs.

902. Hudson, C. A. "Computers in Manufacturing." *Science* 215 (February 12, 1982), 818-825.

Includes a section on robotics, with information on categories of industrial robots, what they are, how they are used, barriers to utilization, and research projects, including applications being studied at the Robotics Laboratory at Westinghouse. Illustrations include drawings of an industrial robot system called APAS (Automated Programmable Assembly System), and the system used to control it.

903. Hudson, Christopher A. "Robotics and the Factory of the Future." *Industrial Development* 152 (September/October 1983), 8-11.

Discusses steps needed to be taken to go from present methods of manufacture to the fully automated factory of the future, using Westinghouse Electric Corporation, Pittsburgh, Pennsylvania, as an example, and describes robots developed by Westinghouse, as well as other industrial robots, and projects to develop automated manufacturing systems.

904. Irwin, Robert D. "Get the Most from Computerized Steel-Collar Workers." *Production Engineering* (August 1981). Reprinted in *Working Safely with Industrial Robots*, pp. 23-27. Edited by Peter M. Strubhar. Dearborn, MI: Robotics International, Society of Manufacturing Engineers, 1986. 266 p.

Discusses types of industrial robots, sales, payoffs of use, safety, and installation. Includes photographs.

905. Johnson, Robert. "Robotics in Manufacturing." *Industrial Education* 73 (May/June 1984), 9-10.

Presents an overview of the robotics industry, industrial robot use, and skill needs of industrial education.

906. Joseph, Earl C. "Robots in Japan Pose Another Challenge to U. S. Industry." *Computerworld* 19 (October 7, 1985), U11, U16.

Reports on production and use of robots in Japan.

907. Kafrissen, Edward, and Stephans, Mark. *Industrial Robots and Robotics*. Reston, VA: Reston Publishing Co., Inc., 1984. 396 p.

Covers microcomputer servo control, actuator, sensor, mechanical, vision, software, and other systems of an industrial robot. Also includes applications, robot work cell environments, and economics and growth of robots. Provides glossaries, lists of manufacturers, bibliography, and other information. Includes photographs, drawings, and schematics.

908. Katzan, Harry, Jr. *A Manager's Guide to Productivity, Quality Circles, and Industrial Robots*. New York: Van Nostrand Reinhold Company, 1985. 136 p.

Covers productivity and quality concepts, quality circle principles, methods and strategic planning, industrial robot principles, and advanced topics in robotics.

909. Kosnett, Jeff, and van Cott, Wendy. "Super Stocks for Tomorrow: Robotics." *Changing Times* 41 (August 1987), 60-64.

Discusses the state of the industrial robotics industry, and industrial robotics companies, particularly International Robomation Intelligence, Robotic Vision Systems, and Medar. Includes color photographs.

910. Krasnoff, Barbara. "The Truth About Robots." *Popular Mechanics* 158 (September 1982), 77-81, 166-167.

Describes industrial robots and how they work. Box reports on hobby robots. Includes color and black-and-white photographs.

911. Lammineur, P., and Cornillie, O. *Industrial Robots*. Oxford, England: Pergamon Press, 1984. 154 p.

Provides drawings and explanatory text for various types of industrial robots.

912. Levin, Bob; Doi, Ayako; and Gibney, Frank, Jr. "Here Come the Robots." *Newsweek* 100 (August 9, 1982), 58-59.

Reports on the robotics industry and industrial applications in Japan. Provides color photographs, including one of an experimental bedside robot for hospitals.

913. Lindorff, Dave. "The Robotics Revolution." *Madison Avenue* 26 (February 1984), 38-39, 42-43.

Reports on ways industrial robots are used and the industrial robot market.

914. Lovedahl, Gerald G. "Robotics: A New Challenge for Industrial Arts." *Man Society Technology* 42 (April 1983), 14-15.

Considers uses of robots in industry, predicted improvements in robots, and changes to come in industry.

915. Lynn, Leonard. "Japanese Robotics: Challenge and - Limited - Exemplar." *Annals. American Academy of Political and Social Science* 470 (November 1983), 16-27.

Discusses use of industrial robots in Japan, Japanese robot industry, and social impact of robotics there.

916. Maher, Philip. "Coming to Grips with the Robot Market." *Industrial Marketing* 67 (January 1982), 93-94, 96, 98.
Discusses the robot market in the United States and Japan. Illustrated with drawing.

917. MaHill, John. "Exploding Population of Automatic Workers." *Technology Review* 82 (February 1980), 78, 80.
Reports on industrial robot use around the world.

918. Mair, Gordon. *Industrial Robotics*. New York: Prentice Hall, 1988. 354 p.
Covers introduction to industrial robots, origin of the robot species, classifying and describing industrial robots, robot design and control, the control unit, electric drives and motor control, fluid power drives, transmission sensors, end effectors, programming, applications, economics and justification, human factors, the automated factory, and future of robotics. Illustrated with photographs and drawings.

919. Malcolm, Douglas R., Jr. *Robotics: An Introduction*. Boston, MA: Breton Publishers, 1985. 368 p.
Covers introduction, basic features of manipulator, controllers, robotic programming, operational aids, hydraulic and pneumatic drive systems, DC and AC motor operation, servo system control, robotic gears and linkages, interfacing, end effectors, robotic sensors, applications, and growth and costs. Illustrated with photographs and drawings.

920. Malcolm, Douglas R. *Robotics*. 2nd ed. Boston, MA: PWS-Kent Publishing Company, 1988. 395 p.
Updated edition. See number 919.

921. Marbach, William D.; Lampert, Hope; and Cook, William J. "The Factory of the Future." *Newsweek* 100 (September 6, 1982), 69.

Reports on the use of computers and robots in industry. Includes photographs.

922. Marsh, Peter. "Britain Grapples with Robots." *New Scientist* 86 (April 24, 1980), 183-187.

Reports on the use of industrial robots in England, provides facts about robots, and discusses worker's fear of robots and the importance of robot flexibility. Includes photographs. See also Peter Marsh's comments on Britain's robots on page 178 of this issue.

923. Marsh, Peter. *The Robot Age*. London: Sphere Books, Inc., 1982. 183 p.

Considers history of machines and robots, present state of industrial robots, and industrial robots of the future, including factories in space and self-replicating robots. Illustrated with photographs and drawings.

924. Masterson, James W.; Poe, Elmer C.; and Fardo, Stephen W. *Robotics*. Reston, VA: Reston Publishing Company, Inc., 1985. 263 p.

Covers what a robot is, historical industrial robot development, basic characteristics, fundamentals and robotic features, factors to consider in selection and implementation, automated manufacturing systems, sensing systems, fluid power and electric machinery and power systems, nonintelligent, intelligent, and very intelligent controllers, and computers in control. An appendix provides electrical/electronic symbols and fluid power symbols. Illustrated with photographs and drawings.

925. Maus, Rex, and Allsup, Randall. *Robotics: A Manager's Guide*. New York: John Wiley & Sons, Inc., 1986. 238 p.

Covers the manufacturing environment, computer automation, robots and the Japanese, robot design, fabrication, assembly, inspection, materials handling, flexible manufacturing systems, economics, and the future. Includes glossary, reference tables, and lists of robot organizations and manufacturers. Illustrated with photographs and drawings.

926. Miller, Rex. *Fundamentals of Industrial Robots and Robotics*. Boston: PWS-Kent Publishing Company, 1988. 288 p.
Covers an introduction to robots, types of robots, industrial robots and their drive systems, robot sensors, manipulator control methods, the robot and the computer, and uses for robots. Includes lists of manufacturers and equipment, spec sheets, and a cross comparison of robots. Illustrated with photographs and drawings.

927. Mitchell, Roger H., and Mabert, Vincent A. "Robotics for Smaller Manufacturers: Myths and Realities." *Business Horizons* 29 (July/August 1986), 9-16.
Describes the industrial robot, its components and characteristics, discusses current robot applications, areas of potential applications, and why to use robots, and gives suggestions on implementation. Illustrated with drawings.

928. Mitchell, Russell. "Boldly Going Where No Robot Has Gone Before." *Business Week* no. 2978 (December 22, 1986), 45.
Reports on uses of robots in industry and on industrial robot manufacturers. Includes color photograph.

929. Morgan, Chris. *Robots; Planning and Implementation*. London & Berlin: IFS (Publications) Ltd. & Springer-Verlag, 1984. 195 p.
Discusses defining the need for a robot, choosing a robot, specifying, ordering, and installing the robot system, safety and robot systems, informing and training the workforce, and finances. Provides case studies. Includes photographs and diagrams.

930. Mortimer, John, ed. *Decade of Robotics*. London & Berlin: IFS (Publications) Ltd. & Springer-Verlag, 1983. 168 p.
A glossy magazine format compilation of articles and advertisements featuring, primarily, industrial robots. Covers the robot experience, robot design, robots in action, robot management, robots on the fringe (underseas, mining, in space, and the robot industry in Japan), and the future. Includes a directory of the world's robot makers and

suppliers. Illustrated with color and black-and-white photographs, and drawings.

931. Nagler, Ben. "Off-Line Programming: The Next Big Step in Robot Technology." *Computerworld* 19 (October 7, 1985), U13, U16.

Describes a method of programming industrial robots away from the workplace at a remote terminal.

932. Norman, Colin. "The New Industrial Revolution." *The Futurist* 15 (February 1981), 30-34, 36-40.

Includes report on the use of robots in industry. Illustrated with photographs, including pictures of industrial robots.

933. Obrzut, John J. "Robotics Extends a Helping Hand." *Iron Age* 225 (March 19, 1982), 59, 61-62, 64-65, 68, 70-71, 74-76, 78, 80-83.

Discusses the robotics industry, provides a chart, map, and graphs on industrial robots around the world, considers industrial robot applications and management resistance to robots, and provides list of robot suppliers, robotic accessory producers, robotics service companies, and a glossary. Illustrated with photographs. Cover of this issue shows an industrial robot arm.

934. O'Neill, Gerard K. "Robots Who Reproduce." *Across the Board* 21 (June 1984), 30-31, 33-34, 36-39.

Discusses uses and economics of robots, robotic sensory perception, effects of the automated factory, and how the United States can catch up with other countries. Illustrated with color photograph of Fanuc factory floor in Japan.

935. Organisation for Economic Cooperation and Development. *Micro-Electronics, Robotics and Jobs*. Paris: Organisation for Economic Cooperation and Development, 1982. 249 p. ERIC document ED 229 641.

Includes Volkholz, Volker, "Trends in the Use of Industrial Robots in the 80s - the Case of the Federal

Republic of Germany," p. 173-196; Carlsson, Jan, and Selg, Hakan, "The Promotion of Robotics and CAD/CAM in Sweden," p. 197-233; Yonemoto, Kanji, "Robotisation in Japanese Industries," p. 235-245; and Le Quement, Joël, "Robotics An Industrial Challenge," p. 247-263.

936. Osborne, David M. *Robots: The Application of Robots to Practical Work.* Detroit, MI: Midwest Sci-Tech Publishers, Inc., 1984. 232 p.

Covers all aspects of industrial robot use, and includes chapters on nonindustrial robots, medical uses of very small robots within the human body, uses of very large robots, and transportation robots. Illustrated with drawings.

937. Ottinger, Lester V. "Robotics, Part 1: Robotics for the IE: Terminology, Types of Robots." *Industrial Engineering* 13 (November 1981), 28-35.

Covers robot terminology, work envelopes, robot classifications, jobs robots can do, and major robot vendors. Includes drawings. Cover of this issue includes a color drawing of an industrial robot.

938. Ottinger, Lester V. "A Plant Search for Possible Robot Applications." *Industrial Engineering* 13 (December 1981), 26-32.

Describes how to survey manufacturing facility for robot applications, lists typical robot tasks, technical and economic considerations, and human factors, and provides a Part Handling Application Data Sheet form and Impressions Regarding Robot Applications form. Illustrated with photographs.

939. Ottinger, Lester V. "Evaluating Potential Robot Applications in a System Context." *Industrial Engineering* 14 (January 1982), 80-87.

Discusses choosing an industrial robot based on application evaluation. Illustrated with photographs and drawings.

940. Ottinger, Lester V. "Engineering Robot Systems for Existing Facilities." *Industrial Engineering* 14 (February 1982), 20-23.
Discusses task analysis, equipment specification, the robot hand, feedback devices, parts presentation, floor plans, safety, and maintenance of industrial robots. Illustrated with photograph.

941. Ottinger, Lester V. "Engineering a Robot System for a New Facility." *Industrial Engineering* 14 (April 1982), 86-90.
Discusses facilities, tasks, and environmental needs of industrial robots. Illustrated with photographs and drawing.

942. Ottinger, Lester V. "Fabrication and Installation of Robot Systems." *Industrial Engineering* 14 (May 1982), 92-94, 96, 98.
Discusses personnel concerns, and installation of industrial robots. Illustrated with photographs.

943. Ottinger, Lester V. "Robot System's Success Based on Maintenance." *Industrial Engineering* 14 (June 1982), 38-40, 42-43.
Discusses maintenance program for industrial robots. Illustrated with photograph.

944. Ottinger, Lester V. "Robot Implications for Production and Inventory Control." *Industrial Engineering* 14 (July 1982), 16-22.
Considers implications of an industrial robot system on production control and inventory. Illustrated with drawings.

945. Ottinger, Lester V. "Questions Potential Robot Users Commonly Ask." *Industrial Engineering* 14 (August 1982), 28-30, 33.
Considers questions involving use of industrial robots. Illustrated with photograph.

946. Ottinger, Lester V. "Robots and Other Technologies in the Automated Factory." *Industrial Engineering* 14 (September 1982), 26-32.

Discusses considerations in industrial robot use. Illustrated with photographs.

947. Owen, Tony. *Assembly with Robots*. Englewood Cliffs, NY: Prentice-Hall, Inc., 1985. 211 p.

Covers all aspects of assembly with robots. Also considers why to use robots in assembly operations, safety procedures, evaluation of a robot system, economics of alternative systems, and the future. Includes photographs and diagrams.

948. Paul, Richard. "Robots, Models, and Automation." *Computer* 12 (July 1979), 19-27.

Describes technical developments which led to development of the industrial robot, sensor-controlled robots, the use of a camera for robot vision, and problems of vision and sensory control in assembly of products. Includes photographs.

949. Productivity International, Inc. *A Survey of Industrial Robots*. 2nd ed. Dallas, TX: Leading Edge Publishing, Inc., 1982. 338 p.

Considers pros and cons of robot applications, user acceptance, fundamentals of industrial robots, implementing a robot system, economic justification, integration guidelines, applications, and the future. Bulk of book provides tables of information on various models of industrial robots. Includes photographs.

950. Reed, David. "Robots March on U. S. Industry." *Reader's Digest* 126 (April 1985), 188-192.

Reports on industrial robots, applications, and possible effects on job availability.

951. Reed, Dean, and Harden, Thomas K. *Industrial Robots*. 1986. 50 p. ERIC document ED 266 937.

Discusses history and status of robot industry, noneconomic factors related to robot use, labor considerations, funding, robot applications, and future prospects.

952. Reeve, R. C. "Where Are the Robots?" *Computerworld* 19 (June 10, 1985), ID25-ID26, ID30-ID32.
Discusses reasons why there are more industrial robots used in Japan than in the United States. Illustrated with photograph, charts, and graphs.

953. Rehg, James A. *Introduction to Robotics*. Englewood Cliffs, NJ: Prentice-Hall, Inc., 1985. 230 p.
Covers introduction to industrial robotics, history, robot classification, end-of-arm tooling, robot and controller operation, sensors and interfacing, programming, safety, and human interface. Includes photographs and drawings.

954. Rifkin, Glenn. "A Look at Robotics in the U. S." *Computerworld* 19 (1985), U2-U4, U6, U8, U10, U12, U14-U15.
Discusses industrial robot use and provides brief history of robots. Illustrated with photographs.

955. Robotics International, Society of Manufacturing Engineers. *Robotics Today; '82 Annual Edition*. Dearborn, MI: Robotics International, Society of Manufacturing Engineers, 1982. 369 p.
A collection of articles from *Robotics Today* from Summer 1979 to Fall 1981. Covers the basics of industrial robots, human factors, machine loading/material handling, welding, assembly, casting and forging, painting, other applications, implementation, and vision and sensing. Includes directory of manufacturers and distributors. Illustrated with photographs and drawings.

956. Rosenblatt, Jean. "The Robot Revolution." *Editorial Research Reports* 1 (May 7, 1982), 345-363.
Covers robots in history, technological advances in robotics, United States - Japanese competition, and

consequences for society. Illustrated with drawings and photograph.

957. Salim, Parvez. "The Robots Are Coming." *Risk Management* 31 (May 1984), 30-33, 35-36, 38.

Covers robots in retrospect, what an industrial robot is, types, applications, future implications, hazards, and economics of using robots. Illustrated with drawings.

958. Schefter, Jim. "New Workers on the Assembly Line: Robots That Think." *Popular Science* 216 (June 1980), 46, 48, 50, 53-54, 57.

Describes research on improving industrial robots, and uses of robots in industry. Deals particularly with Unimation's PUMA. Includes photographs and drawings.

959. Seering, Warren P. "Who Said Robots Should Work Like People?" *Technology Review* 88 (April 1985), 58-67.

Discusses tendency to build robots to act like humans rather than redesign products so they can be assembled by robots that act like machines and not like humans, manufacturing tasks robots do well that human beings are not particularly good at, and the difference between Japanese simple task robots and United States general task robots.

960. Stefanides, E. J. "Lending a Helping (Mechanical) Hand." *Design News* 43 (February 9, 1987), 130, 133.

Describes research, especially on industrial robots, being conducted at the Robotics Institute at Carnegie Mellon University, Pittsburgh, Pennsylvania. Illustrated with color photographs.

961. Susnjara, Ken. *A Manager's Guide to Industrial Robots*. Englewood Cliffs, NJ: Prentice-Hall, Inc., 1982. 186 p.

Includes information on what an industrial robot is, applications, installation, economics, human relations, the automatic factory, and the future. Provides glossary. Includes photographs.

962. Tanner, William R., ed. *Industrial Robots*. Vol. 1: *Fundamentals*. Dearborn, MI: Robotics International of the Society of Manufacturing Engineers, 1981. 446 p.

Covers basics, implementation, justification, human factors, capabilities, applications, advanced technology, manufacturers and distributors, and organizations. Includes photographs.

963. Tanner, William R., ed. *Industrial Robots*. Vol. 2: *Applications*. Dearborn, MI: Robotics International of the Society of Manufacturing Engineers, 1981. 484 p.

Covers applications in material handling, machine loading, die casting, investment casting, press loading, forging and heat treating, foundry, plastic molding, welding, machining, finishing, inspection, and assembly.

964. Teresko, John. "Fallout Continues." *Industry Week* 221 (June 25, 1984), 91, 93, 95.

Reports on status of industrial robot market and specific companies.

965. Teresko, John. "Ready to Move? Industry's Shipments and Backlogs Growing." *Industry Week* 225 (June 24, 1985), 28.

Reports on industrial robot market.

966. Teresko, John. "Robots Come of Age." *Industry Week* 212 (January 25, 1982), 35-39, 42.

Discusses the impact of industrial robots, and reports on robot manufacturers, and on a roundtable discussion of robotics sponsored by the American Management Association. Includes profile of George C. Devol, one of the founders of Unimation, the world's first industrial robot manufacturer. Includes color and black-and-white photographs. Cover of this issue shows a color drawing of a futuristic robot.

967. Teresko, John. "The Robots Are Ready; But Where Are the Factories of the Future?" *Industry Week* 223 (October 23, 1984), 42-44, 48.

Reports on industrial robot market, specific companies, applications, and future uses. Illustrated with photographs.

968. Thackray, John. "America's Robot Rising." *Management Today* (February 1983), 54-57, 121.
Discusses industrial robot market, applications, blue-collar job loss, and the Japanese robotics industry.

969. Todd, D. J. *Fundamentals of Robot Technology*. New York: John Wiley & Sons, 1986. 244 p.
Covers introduction, geometric configurations for robots, operation, programming and control, actuators, sensors, performance specifications, applications, teleoperated arms, mobile robots, automated guided vehicles, robotics and artificial intelligence, and economic and social aspects. Illustrated with photographs and drawings.

970. Toepperwein, L. L.; Blackmon, M. T.; Park, W. T.; Tanner, W. R.; and Adolfson, W. F. *Robotics Applications for Industry; A Practical Guide*. Park Ridge, NJ: Noyes Data Corporation, 1983. 326 p.
Covers robot configuration, sensors, tooling, and workstation integration. Includes section on the use of robots in motor vehicle manufacture, classified glossary, and bibliography. Illustrated with drawings.

971. Ullrich, Robert A. *The Robotics Primer*. Englewood Cliffs, NJ: Prentice-Hall, Inc., 1983. 121 p.
Discusses what a robot is, robots at work, economics and robots, research in robotics, and robots and organizations. Provides annotated bibliography.

972. Wetherbe, James C. and Conrad, Scott. "What MIS Executives Need to Know About Robotics." *Journal of Systems Management* 34 (May 1983), 38-42.
Describes industrial robots and the robotics marketplace, discusses trends, cost, uses, vision and intelligence, job replacement, and union reactions, and

provides recommendations to management information system managers. Illustrated with photographs.

973. Whitney, David E. "Real Robots Do Need Jigs." *Harvard Business Review* 64 (May/June 1986), 110-116.
Discusses problems involving design of industrial robots, and the fallacy of thinking that robots are mechanical people. Includes drawings.

974. Wiegner, Kathleen K. "The Dawn of Battle." *Forbes* 128 (October 26, 1981), 77, 79.
Reports on the industrial robotics industry in Japan, with special emphasis on Fujitsu Fanuc's factory, run mostly by robots. Includes photograph.

975. Wolfe, Gene. "Robotics: Living with the Iron Collar Worker." *Plant Engineering* 37 (February 3, 1983), 51-54.
Answers questions about industrial robot use, such as how good applications for robots can be identified, will robots put people out of work, are robots dangerous machines, and so forth. Illustrated with photographs.

Robot Manufacturers

976. "A Computer Behind the Wheel." *High Technology Business* 8 (January 1988), 8.
Describes a self-guided vehicle manufactured by Caterpillar Industrial to help with such tasks as moving heavy parts to production machines, and as a replacement for fork-lift trucks. Illustrated with photograph.

977. "A Nuts-and-Bolts Interview with 'Father' of Robotics." *Computerworld* 17 (April 25, 1983), 15.
Reports on an interview with Joseph F. Engelberger, cofounder of Unimation, Incorporated. Illustrated with photograph.

978. "Automatix: Gaining an Edge with Brainy Robots." *Business Week* no. 2847 (June 18, 1984), 60, 64.

Reports on industrial robot and artificial vision system manufacturer, Automatix, Incorporated, of Billeria, Massachusetts.

979. "British Robots. Japanese Style." *The Economist* 280 (September 26, 1981), 69-70.

Reports on industrial robot makers in Great Britain.

980. "Design with the Robot in Mind." *Design News* 44 (March 21, 1988), 46.

An interview with Vincent M. Altamuro, president of Robotics Research Consultants, Toms River, New Jersey, discussing what robots can do and listing tips for product design so that robots can assemble products more easily. Illustrated with photograph.

981. "Fanuc Edges Closer to a Robot-Run Plant." *Business Week* no. 2664 (November 24, 1980), 56.

Reports on Fujitsu Fanuc Limited's plant in Japan that uses robots to build other robots and computerized tools, operating 24 hours a day, with humans assembling robot and machine-made parts for eight hours a day. Includes drawing.

982. "Government Scheme Could Give Robots a Hand." *New Scientist* 85 (January 31, 1980), 320.

Summarizes an interview with Doug Hall, managing director of Hall Automation, a robot manufacturer in England. Outlines five target areas for robot applications: assembly, working in small spaces, forging, die casting, and spraying glass fibre. Includes photograph.

983. "How Robots Could Help Rev up Caterpillar." *Business Week* no. 2852 (July 23, 1984), 73.

Reports on Caterpillar Tractor Company and Advanced Robotics Corporation of Columbus, Ohio, and work on arc-welding robots. Includes color drawing.

984. "Racing to Breed the Next Generation." *Business Week* no. 2640 (June 9, 1980), 73, 76.
Reports on developments in industrial robot design and applications for the next generation of robots. Includes photograph.

985. "Renault's Robots." *Popular Mechanics* 156 (October 1981), 193.
Describes industrial robots made and used by Renault in its automotive plants in France. Includes photographs.

986. "Robot Automates Adhesives Dispensing." *Design News* 44 (September 19, 1988), 28.
Describes a robot that can lay down a consistent bead of adhesive on a part for assembly, developed by CRS Plus Automation, Burlington, Ontario, Canada.

987. "Robot Raid." *Economist* 285 (December 11, 1982), 83.
Reports on takeover of Unimation by Westinghouse Electric. Illustrated with photograph.

988. "Robots at the Ready: Cincinnati Milacron Looks Beyond the Slump." *Fortune* 103 (June 1, 1981), 9.
Reports on industrial robot market in 1981, particularly that of Cincinnati Milacron. Includes color photograph of James A. D. Geier, president of the company.

989. "Vendors of Robots." *Datamation* 27 (January 1981), 99-100, 104.
Provides descriptions and sources for industrial robots offered by 18 manufacturers. Includes photographs of some robots.

990. Alpert, William M. "Rout of the Robots." *Barron's* 64 (November 5, 1984), 15, 51.
Reports on a number of industrial robot manufacturers. Illustrated with photographs and drawing.

991. Bak, David J. "This Robot Sees!" *Design News* 44 (February 15, 1988), 204-205.

Describes a robot/vision system for use in microelectronics inspection/assembly situations, developed by MicroRobotics Systems, Incorporated, Chelmsford, Massachusetts. Illustrated with photograph and drawing.

992. Banks, Howard. "Entrepreneurial Capitalism." *Forbes* 140 (July 13, 1987), 228.

Reports on the Italian firm Bisiach and Carru SpA, maker of precision robots for welding vehicle bodies. Includes color photograph.

993. Bylinsky, Gene. "Japan's Robot King Wins Again." *Fortune* 115 (May 25, 1987), 53-54, 56, 58.

Reports on Fanuc, Limited, producer and user of industrial robots. Includes color photographs.

994. Dizard, John W. "Giant Footsteps at Unimation's Back." *Fortune* 105 (May 17, 1982), 94-99.

Reports on developments at the industrial robotics firm Unimation, Incorporated. Includes color photographs.

995. Fromson, Brett Duval. "Adept Technology Inc." *Fortune* 116 (August 3, 1987), 114.

Reports on Adept Technology robot used for assembling small parts. Includes color photograph of the robot building a circuit board.

996. Hall, Douglas. "My Life with Robots." *New Scientist* 97 (February 17, 1983), 468-470.

Traces the development of industrial robots in Great Britain from 1966 through the 1970's, and the rise and assimilation of Hall Automation, one of the first industrial robot manufacturers. Includes photographs.

997. Krasnoff, Barbara. "Programming Factory Robots with PCs." *PC Magazine* 3 (August 21, 1984), 120-122, 125, 127.

Reports on industrial robot - microcomputer systems developed by International Business Machines (IBM). Illustrated with photographs.

998. Marsh, Peter. "Electrical Giant Switches on to Foreign Robots." *New Scientist* 96 (October 7, 1982), 23.
Reports on industrial robots being developed at GEC Electrical Products Company in Rugby, England. Includes photograph.

999. McCormick, Jay. "Overblown, Oversold, Over-Everythinged." *Forbes* 135 (March 25, 1985), 191, 194.
Reports on Prab Robots, Incorporated, of Kalamazoo, Michigan, and specific uses of the company's robots, such as to bake bread for Kroger, move empty face cream bottles for Mary Kay Cosmetics, and dismantle World War II shells for the military. Includes photograph.

1000. Phillips, Leonard A. "The Robot Man." *Technology Review* 85 (May/June 1982), 37.
Chronicles the life and work of George Devol, pioneer industrial robot designer of Unimation, Incorporated, whose patents were the basis for the industrial robot industry. Illustrated with portrait drawing.

1001. Saporito, Bill. "IBM's No-Hands Assembly Line." *Fortune* 114 (September 15, 1986), 105, 108-109.
Reports on the IBM PC convertible, laptop microcomputer and industrial robots it can control. Includes color photographs of an industrial robot.

1002. Solomon, Stephen D. "Pittsburgh." *Inc.* 7 (January 1986), 58-61, 64, 66.
Relates the history of American Robot Corporation. Illustrated with photograph.

1003. Thackray, John. "The Great American Robot Fiasco." *Management Today* (November 1986), 86-88, 92.

Reports on industrial robot market, manufacturers, and applications. Illustrated with photographs.

1004. Tracy, Eleanor Johnson. "How to Sell Robots in Robotland." *Fortune* 110 (August 6, 1984), 60.
Reports on American Robot Corporation and its industrial robot sales in Japan. Includes color photographs.

1005. Yasaki, E. K. "Japanese Push Robotics." *Datamation* 27 (July 1981), 56, 61.
Reports on a Japanese government program, a leasing company to help domestic producers of industrial robots.

Robot Uses and Users

1006. "Champion Spark Plug Uses Robot in Manufacturing Operation." *Aviation Week/Space Technology* 122 (April 29, 1985), 221.
Describes use of robot to clean cement residue from holes in spark plugs. Includes photograph.

1007. "GE Dishwasher: Automation Makes It a Just-in-Time Plant." *Plastics World* 42 (September 1984), 56-57.
Describes use of robots to build dishwashers.

1008. "How Robots Are Already Cutting Costs for GE." *Business Week* no. 2640 (June 9, 1980), 68.
Reports on how General Electric Company uses industrial robots. Includes photograph.

1009. "Look, No Hands; Brave New World at the Factory." *Time* 118 (November 16, 1981), 127.
Describes the Yamazaki Machinery Works plant in Nagoya, Japan, which is run mostly by robots. Includes photograph.

1010. "Machine Vision: Communications Link for Manufacturing." *Industrial Education* 77 (76) (September 1987), 9.

Describes vision/robot system used to perform testing and aligning of circuit packs used in radio frequency products. Illustrated with photograph.

1011. "Robots at Work: Todd Pacific Shipyards Corporation." *Industrial Engineering* 15 (May 1983), 58-62.

Describes use of welding robots in shipbuilding. Illustrated with photographs.

1012. "Robots Get a Warm Welcome in Kentucky." *Business Week* no. 2735 (April 19, 1982), 48-49.

Reports on plans by Yamazaki Machinery Works, Limited, for a fully automated Japanese machine tool factory run by robots and employing only six people, in Florence, Kentucky. Includes photograph.

1013. "Touch and Go for Robots." *Science News* 127 (March 23, 1985), 187.

Reports on projects to develop a sense of touch for industrial robot "hands" or grippers.

1014. "Welding Robots Made Cost-Effective." *Industrial Education* 75 (September 1986), 42-44.

Reports on uses of welding robots. Illustrated with photographs.

1015. "Westinghouse Uses Robot Gripper to Assemble Microwave Components." *Aviation Week/Space Technology* 126 (March 2, 1987), 81.

Describes robot gripper used in assembly. Illustrated with color photographs.

1016. Brown, Stuart F. "Building Cars with Machines That See." *Popular Science* 227 (October 1985), 86-88, 126.

Describes applications of industrial robots equipped with machine vision. Includes color photographs and drawings.

1017. Bylinsky, Gene. "The Race to the Automatic Factory." *Fortune* 107 (February 21, 1983), 52-60, 64.

Reports on progress in industry with flexible manufacturing systems, which integrate computer-controlled machinery centers, robots, remotely-guided carts, and other computer-related elements into an automated factory that could become nearly workerless. Includes color photographs and a drawing of components of the system. Cover of this issue shows a color drawing of industrial robots.

1018. Corrado, Joseph K. "Keyboard Is Engineered for People and Robots." *Design News* 40 (January 23, 1984), 110-111.

Describes keyboard designed for fast assembly by robots and easy use by humans, developed by Digitran Company, Pasadena, California. Illustrated with photographs and drawing.

1019. Corrado, Joseph K. "Laser Power Is Placed Under Robotic Control." *Design News* 40 (December 3, 1984), 94-96.

Describes robot use to control laser tools. Illustrated with photographs and drawings.

1020. Costlow, Terry. "Robot Loads Bases, Lead Frames in Boats." *Electronics* 54 (October 20, 1981), 44.

Describes a robot that can load 6500 pieces an hour, in a job in which six people can load 6000 pieces an hour. Includes photograph.

1021. Darrow, Barbara. "New Turf for Robots." *Design News* 43 (March 23, 1987), 56-58, 60, 61.

Describes ways robots are being used in industry, such as coating freezer coils, making cellular phone connections, in apparel manufacture, and for fruit picking. Illustrated with color photographs and drawing.

1022. Dizard, John W. "Machines That See Look for a Market." *Fortune* 110 (September 17, 1984), 87-88, 92, 96, 100, 104.

Reports on the use of machine vision in manufacturing, including use with industrial robots, especially by General Motors. Includes color photographs.

1023. Dillon, Thomas F. "Behind the Rise in Industrial Robot Use." *Purchasing* 92 (February 11, 1982), 78-80.

Describes what a robot is, industrial robot classifications, and uses of industrial robots. Includes color photographs.

1024. Dodd, John. "Robots: the New 'Steel Collar' Workers." *Personnel Journal* 60 (September 1981), 688-695.

Describes history of industrial robots, who uses them, importance to industry, advantages and disadvantages, costs to employees, union actions, personnel manager concerns, communication importance, regulating rate of introduction, and conclusions. Includes photographs and drawing.

1025. English, Carey W. "Factories That Turn Nuts into Bolts." *U. S. News & World Report* 101 (July 14, 1986), 44-45.

Discusses flexible manufacturing systems, including the use of robots in plants using these systems. Includes color photographs and drawings.

1026. Halberstam, David. "Robots Enter Our Lives." *Parade Magazine* (April 10, 1983), 16-17, 19.

Reports on ways robots are used in industry. Illustrated with photographs.

1027. Hampton, William J. "GM Bets an Arm and a Leg on a People-Free Plant." *Business Week* no. 3069 (September 12, 1988), 72-73.

Describes Saginaw Vanguard plant in Saginaw, Michigan, a fully automated factory where experimental robots deliver parts and assemble front-wheel drive axles for General Motors. Includes color photographs.

1028. Jungreis, Abigail. "Revolution in the Workplace." *Scholastic Update* (Teacher's Edition) 116 (September 2, 1983), 17-19.

Reports on use of robots in industry, and computers in offices. Includes color photographs.

1029. Keller, Erik L. "Robot Tunes Mobile Radios." *Electronics* 56 (April 7, 1983), 48-49.
Describes the use of a robot to tune radios at General Electric Company, Lynchburg, Pennsylvania.

1030. Klass, Philip J. "Semi-Automated Plant Enhances Quality, Cuts Manufacturing Cost." *Aviation Week & Space Technology* 121 (November 26, 1984), 66-68, 73.
Describes Westinghouse's use of robots to manufacture printed circuit board modules. Includes photographs.

1031. Kozlov, Alex. "The Automation Generation." *Omni* 11 (March 1989), 18, 22.
Comments of computer scientist David Bourne, mechanical engineer Paul Wright, master machinist Jim Dillinger, and master machinist Dan McKeel regarding use of industrial robots now and in the future. Illustrated with color drawing.

1032. Lawson, James; Reidy, John; Renner, G. Frederick; Rosen, Karen; and Smolak, Lise. "Technology for the Factory of the Future." *Annals. American Academy of Political and Social Science* 470 (November 1983), 56-67.
Discusses current and future uses of robots in industry.

1033. Petre, Peter. "How GE Bobbled the Factory of the Future." *Fortune* 112 (November 11, 1985), 52-54, 58, 62-63.
Describes General Electric's use of automation in manufacturing and includes discussion of their industrial robot manufacture and use. Includes color photographs.

1034. Port, Otis. "This Robot Makes Chips in a Clean Room of Its Own." *Business Week* no. 3037 (February 8, 1988), 57.
Reports on a robot that can produce microchips in a cylindrical vacuum chamber, being developed at the Center for Robotic Systems in Microelectronics at the University of

for Robotic Systems in Microelectronics at the University of California at Santa Barbara. Illustrated with color drawing.

1035. Quinlan, Joe. "Rolling Robots Move In; Guided Vehicles with Smart Controls Close Gap in Factory Automation." *High Technology* 3 (January 1983), 26-27.

Describes rolling robots called automatic guided vehicle systems (AGVS), and how they are controlled. Includes photographs and drawing.

1036. Rathmill, Keith, ed. *Robotic Assembly*. Bedford, England & Berlin: IFS (Publications), Ltd. & Springer-Verlag, 1985. 365 p.

Includes papers covering the areas of assembly robots, robot assembly systems and applications, product design for robot assembly, programming systems, sensory systems, and the economics of using assembly robots. Well illustrated with photographs.

1037. Scott, David. "Robot Auto Inspector." *Popular Science* 230 (February 1987), 66.

Reports on the Vision Inspection and Calibration Test by Robot (VICTER), a robot used for quality-control inspection in car production installations.

1038. Shulman, H. G., and Olex, M. B. "Designing the User-Friendly Robot, a Case History." *Human Factors* 27 (February 1985), 91-98.

Describes the design of user interfaces for an industrial spray-painting robot. Illustrated with photographs and schematic.

1039. Verney, Thomas P.; Hollon, Charles J.; and Rogol, George. "HR Planning for Robots in the Work Place." *Personnel* 63 (February 1986), 8-9.

Discusses results of a survey of large industrial firms regarding their planned future use of industrial robots. Illustrated with cartoon accompanied by robot blues poem-song.

1040. Wadhwani, Romesh T. "Integrating Robot Power into Automated Factory Systems." *Management Review* 73 (June 1984), 8-14.

Discusses the fitting of robots into a computer-integrated manufacturing system. Illustrated with photographs.

1041. Whiteside, David; Brandt, Richard; and Schiller, Zachary. "How GM's Saturn Could Run Rings Around Old-Style Carmakers." *Business Week* no. 2878 (January 28, 1985), 126, 128.

Includes description of how robots will be used to build cars at the Saturn Corporation facility. Illustrated with photographs.

OTHER ROBOT USES

Aeronautics, Aircraft Industry, Remotely Piloted Vehicles, and Space Exploration Robots

Aircraft Industry and General Aeronautics Robots

1042. "Air Force Plans to Automate Aircraft Maintenance Depots." *Aviation Week/Space Technology* 125 (December 22, 1986), 55.

Includes description of how robots will be used to repair aircraft. Cover of this issue shows a color picture of Westinghouse industrial robot used for assembly.

1043. "Automated Plate Builder." *Popular Mechanics* 164 (August 1987), 96.

Reports on robotic carousel head used to fasten airplane wings together. Illustrated with color photograph.

1044. "Automatic Spray Painting Device in Operation at Fairchild Republic." *Aviation Week & Space Technology* 117 (August 2, 1982), 78-79.

Reports on the use of a robot to spray paint aircraft parts at Fairchild Republic Company. Includes photographs.

1045. "British Aerospace Installs New Drilling, Fastening Equipment." *Aviation Week & Space Technology* 117 (August 2, 1982), 93.

Reports on robot uses by British Aerospace (company). Includes photographs.

1046. "Composite Inspection System Uses Robot." *Aviation Week/Space Technology* 122 (June 17, 1985), 125.

Describes robot used for panel inspection by Boeing Military Airplane Company. Includes photographs.

1047. "Flying High at General Dynamics." *Dun's Business Month* 123 (February 1984), Special Report, p. X.

Reports on the use of robots to built the F-16 fighter plane, as well as other uses at the General Dynamics plant. Includes photograph.

1048. "Introduction of Robotics Gains European Usage." *Aviation Week & Space Technology* 117 (August 2, 1982), 94, 99, 101.

Reports on uses of robots by European aerospace manufacturers. Includes photographs.

1049. "Japan Air Lines Tests Robotic Spray Coating of Fan Disk." *Aviation Week/Space Technology* 122 (May 13, 1985), 85.

Describes test to use robot for plasma spraying. Includes photograph.

1050. "Northrop Using Robot in F-A-18 Production Work." *Aviation Week & Space Technology* 120 (June 18, 1984), 93.

Describes use of robot to deburr aircraft parts. Includes photograph.

1051. "Robotic Device May Have Applications in Aerospace." *Aviation Week & Space Technology* 118 (April 4, 1983), 59.

Reports on a robotic capability demonstration of Odex I (Odetics, Incorporated, Anaheim, California), with potential applications in aircraft sentry duty and planetary surface exploration.

1052. "Robots Enhance Drilling Productivity." *Aviation Week & Space Technology* 117 (August 2, 1982), 77-79.

Reports on uses of robots by General Dynamics in the production of F-16 aircraft. Includes photograph.

1053. "'Smart' Robot Finds, Removes Aircraft Rivets." *Design News* 40 (January 23, 1984), 17-18.

Describes robot that can distinguish between rivet and skin of aircraft wing or tail, under development at Southwest Research Institute, San Antonio, Texas. Illustrated with drawing.

1054. "USAF Installs Blade Inspection System." *Aviation Week & Space Technology* 117 (August 2, 1982), 63, 65.

Reports on uses of robots and computers in product inspection in the aerospace industry. Includes photograph.

1055. "USAF Looks into Using Robots for Aircraft Maintenance." *Research & Development* 28 (October 1986), 48, 50.

Reports on United States Air Force research program evaluating possible robot replacements for humans performing airplane "turnaround" activities under hazardous conditions, such as during chemical or biological warfare, and eventual robot use in nonmilitary applications, including servicing passenger airliners.

1056. "Vision Sensors Expanding Industrial Robot Flexibility. *Aviation Week & Space Technology* 118 (May 30, 1983), 139, 141, 143-144, 147.

Reports on applications of robots in aerospace manufacturing, with a list of aerospace manufacturers and their uses or planned uses. Includes photograph.

1057. Dornheim, Michael A. "Airframe Makers Expect Computer Techniques to Cut Overhead Costs." *Aviation Week/Space Technology* 125 (December 22, 1986), 56-59.
Includes information on how robots are being used in the aircraft industry. Includes photographs.

1058. Henderson, Breck W. "USAF Expects Robotic Inspection to Cut Maintenance Costs." *Aviation Week/Space Technology* 130 (March 13, 1989), 53-55, 57.
Describes a robotic x-ray and neutron radiography inspection system that can examine a fully assembled aircraft for structural defects. Includes color photographs.

1059. Klass, Philip J. "Circuit Board Assembly Plant to Test Systems." *Aviation Week & Space Technology* 117 (August 2, 1982), 66-67, 71.
Reports on the use of robots and automation in aerospace manufacturing by Westinghouse. Includes photographs. Cover of this issue shows a color photograph of a Westinghouse assembly robot.

1060. Kolcum, Edward H. "C-5A Wing Modification Using Advanced Methods." *Aviation Week & Space Technology* 117 (August 2, 1982), 51-52.
Reports on computer and robot uses in C-5A wing construction. Includes photograph.

1061. Lenorovitz, Jeffrey M. "Sustained Production Rates Hasten Automation in Europe." *Aviation Week/Space Technology* 125 (December 22, 1986), 63, 65-66.
Includes description of how robots are being used in Aerospatial, an aerospace company in France.

1062. Lowndes, Jay C. "Productivity Growth is Key Goal." *Aviation Week & Space Technology* 117 (August 2, 1982), 40-42, 44.
Reports on computer and robot uses in the aerospace industry. Box reports on United States Air Force's program to develop decision-making robots. Illustrated with color and black-and-white photographs.

1063. Lowndes, Jay C. "USAF Seeks Increased Robot Precision." *Aviation Week & Space Technology* 116 (March 1, 1982), 69, 71-72.
Reports on applications of robots in aerospace manufacturing. Includes photograph.

1064. Mordoff, Keith. "Grumman Reduces Sheet Fabrication Costs." *Aviation Week & Space Technology* 117 (August 2, 1982), 74-75.
Reports on uses of robots by Grumman Aerospace Corporation to manufacture aerospace products. Includes photographs and drawing.

1065. O'Lone, Richard G. "Boeing Approaching Use of Robots Cautiously." *Aviation Week & Space Technology* 117 (August 2, 1982), 53, 58-61.
Reports Boeing Company uses of robots in aircraft and missile manufacturing. Includes photographs.

1066. Scott, William B. "New Methods Create Composite Parts." *Aviation Week & Space Technology* 117 (August 2, 1982), 80-81, 83.
Reports on the use of automation and robots to manufacture aircraft structures at Northrop Corporation. Includes photographs.

Remotely Piloted Vehicles (RPV)

1067. "International RPVs and Drones." *Aviation Week/Space Technology* 126 (March 9, 1987), 178.
Provides a chart of information on all remotely piloted vehicles manufactured outside the United States, arranged first by country and then by manufacturer.

1068. "NATO, Industry Groups Study Use of RPVs for Small Ships." *Aviation Week/Space Technology* 127 (September 14, 1987), 95, 99.

Describes a study of vessels suitable as launching and landing pads for remotely piloted vehicles, conducted by the North Atlantic Treaty Organization.

1069. "Robot Spy in the Sky." *New Scientist* 96 (October 28, 1982), 235.
Describes miniature, unmanned helicopter which uses remote control and carries a television camera, developed by ML Aviation of White Waltham, near Maidenhead, England. Illustrated with photograph.

1070. "RPV Packs Latest Recon Tech." *Popular Mechanics* 165 (September 1988), 18.
Reports on Skyeye, remotely piloted vehicle with a camera for nighttime or bad-weather missions. Illustrated with color photograph.

1071. "U. S. RPVs & Drones." *Aviation Week/Space Technology* 126 (March 9, 1987), 176-177.
Provides a chart of information on all remotely piloted vehicles manufactured in the United States, arranged by manufacturer.

1072. Greeley, Brendan M., Jr. "Skepticism, Joint Needs Block Larger Role for RPVs." *Aviation Week/Space Technology* 126 (June 1, 1987), 21-22.
Discusses United States military plans to use remotely piloted vehicles. Includes graph.

1073. Morrocco, John D. "USAF Seeks New RPVs by 1993." *Aviation Week/Space Technology* 126 (February 16, 1987), 17-18.
Discusses the United States Air Force's plans for an unmanned, air reconnaissance system of remotely piloted vehicles.

1074. Schefter, Jim. "Stealthy Robot Planes." *Popular Science* 231 (October 1987), 64-68, 105-106.

Reports on remotely piloted aircraft and their military uses. Illustrated with color photographs and drawings. Cover of this issue shows a remotely piloted aircraft in color.

1075. Taylor, John W. R., and Munson, Kenneth. *Jane's Pocket Book of Remotely Piloted Vehicles: Robot Aircraft Today.* New York: Collier Books, 1977. 239 p.

Includes illustrated history of remotely piloted vehicles (RPV's) and data section with photographs and specific data on all RPV's available at the time of publication.

Space Exploration Robots

1076. "A Different Space Race: Putting Robots on the Planets." *U. S. News & World Report* 69 (November 30, 1970), 45.

Reports on the unmanned lunar vehicle used by the Soviet Union to explore on the moon in 1970. Illustrated with photographs. Of historical value.

1077. "Carnegie-Mellon Team Studies Robotic Technologies for Mars Rover Development." *Aviation Week/Space Technology* 128 (February 8, 1988), 86.

Reports on research on the problems of a self-reliant, autonomous rover that can navigate and collect samples in rugged, soft, unmapped terrain.

1078. "Daddy Longlegs Mars Rover." *Popular Mechanics* 167 (June 1990), 20.

Reports on Ambler, a ten-foot-high Mars rover being developed at Carnegie Mellon University, Pittsburgh, Pennsylvania. Illustrated with drawing.

1079. "Discovery Crew Removed Ice Hazard with Robot Arm." *Aviation Week & Space Technology* 121 (September 10, 1984), 110.

Describes how astronauts on the *Discovery* orbiter used a robot arm to remove ice built up beside a waste-water port which was creating a reentry hazard.

1080. "Helping Robots Slim Down." *USA Today* 116 (June 1988), 13-14.

Reports on research to develop smaller robots for use in space, by Henry Flashner of the University of Southern California, Los Angeles.

1081. "Humans and Robots Cooperating in Space." *USA Today* 114 (June 1986), 14.

Reports on research to develop methods of cooperatively controlling a robot riding in a space shuttle, with the robot doing some of the controlling. Illustrated with drawing.

1082. "Intelligent Robot May Join Space-Station's Crew." *Machine Design* 58 (May 8, 1986), 4.

Describes a seventh "astronaut" that would accompany a six-person crew on a space station, a robot that would evolve through phases such as idiot savant, apprentice, junior partner, and colleague, as it "learned," and as technology became available.

1083. "Moonwalker Relies on Earthly Technology." *Design News* 44 (September 19, 1988), 48.

Describes Skitter, a prototype, three-legged transport vehicle for use on the moon, designed by Jim Brazell, Brice MacLauren, and Gary McMurray at Georgia Institute of Technology, Atlanta.

1084. "NASA Report Urges Developing Robotics, Software for Station." *Aviation Week/Space Technology* 122 (April 22, 1985), 63.

Reports on recommendations of the National Aeronautics and Space Administration Advanced Technology Committee regarding use of robots and artificial intelligence on the space station.

1085. "NASA Robot to Have Private-Sector Spinoffs." *High Technology Business* 9 (March 1989), 37.

Reports on the Flight Telerobotic Service robot to be developed to service the space station *Freedom*.

1086. "NASA's Mighty Robot Riveter." *Popular Mechanics* 165 (July 1988), 17.

Reports on robot riveting machine used to assemble part of the space shuttle at the National Aeronautics and Space Administration's Michoud, Louisiana, assembly facility. Illustrated with color photograph.

1087. "Open the Pod Door." *Space World* W-7-271 (July 1986), 7.

Describes robot designed by Harold Alexander, graduate student at Stanford University, Stanford, California, for use in space.

1088. "Robotic Microscope Could Allow R & D in Space." *High Technology Business* 9 (February 1989), 36-37.

Reports on efforts to adapt a robotic microscope system for use on the space station.

1089. "Robots Put on Overalls and Join the Space Race." *New Scientist* 91 (September 3, 1981), 600.

Reports on robots being developed at Massachusetts Institute of Technology, Cambridge, and the National Aeronautics and Space Administration's Marshall Space Flight Center in Huntsville, Alabama, for use in space projects. Includes photograph.

1090. "Satellite Servicing by Robotic Vehicle Proposed at Meeting." *Aviation Week/Space Technology* 123 (November 25, 1985), 90.

Discusses proposed use of robotic orbital servicing vehicle to increase the life of space craft in geosynchronous orbit.

1091. "Space Robot." *Popular Mechanics* 163 (August 1986), 63.

Describes a proposed space station robot for work outside the space station. Includes photograph.

1092. "Teleoperated Rovers Readied for Space Flights." *Popular Mechanics* 166 (August 1989), 17.

Reports on design being studied for unmanned vehicles for use on missions to Mars and the moon. Illustrated with color drawings.

1093. "Virtual-World Remote Control." *Popular Mechanics* 165 (July 1988), 17.

Reports on a remote control system that lets user explore and interact with a remote site in space, developed by the National Aeronautics and Space Administration's Ames Research Center, Moffett Field, California. Illustrated with color drawing.

1094. "Walking Beam to Rove Mars." *Popular Mechanics* 166 (July 1989), 17.

Reports on walking rover being considered for use on Mars in 1998. Illustrated with drawing.

1095. "What's Astronaut Counterpart?" *Design News* 43 (July 20, 1987), 34-35.

Describes Astro the Astrobaut, a generic robot designed to work outside the space station, developed by McDonnell Douglas and Eagle Engineering, Huntington Beach, California, and reports on other space robot projects. Illustrated with color drawing.

1096. Alexander, Hunter. "Sagdeyev Speaks..." *Space World* Y-2-290 (February 1988), 5.

Reports on views of Roald Zinnurovich Sagdeyev, director of Soviet Union's Space Research Institute, who wants robots rather than humans sent on deep space missions.

1097. Bak, David J. "Three Legs Make Mobile Platform." *Design News* 44 (February 15, 1988), 136-137.

Describes Skitter, a prototype, three-legged transport vehicle for use on the moon, designed by Jim Brazell, Brice MacLauren, and Gary McMurray at Georgia Institute of

Technology, Atlanta. Illustrated with color photograph and drawings.

1098. Beard, Jonathan. "Sweeping up Space Junk." *Discover* 9 (December 1988), 22.

Describes the Autonomous Space Processor Orbital Debris, a robot designed to pick up and dispose of space junk, invented by University of Arizona engineer Kumar Ramohalli. Illustrated with color drawing.

1099. Calder, Nigel. "Earth-Devouring Rocket Ships?" *Science Digest* 92 (March 1984), 93.

Speculates on the idea that self-replicating robot space ships may some day mutate and return to attack Earth. Includes photograph.

1100. Dwiggins, Don. *Robots in the Sky; Explorers of Our Solar System*. San Carlos, CA: Golden Gate Junior Books, 1972. 80 p.

Although dated, provides good coverage of history and explorations of space robots such as the Explorer, Mariner, Viking, and Pioneer series of spacecraft. Illustrated with photographs.

1101. Eberhart, J. "Shuttle 7: We Pick up and Deliver." *Science News* 124 (July 2, 1983), 4-5.

Reports on the uses of a remotely controlled robotic arm on Space Shuttle Flight 7, in June 1983. Illustrated with photographs.

1102. Fennell, Tom, et al. "Buying Canada's Place in Space." *Maclean's* 101 (May 9, 1988), 40-42.

Discusses Canada's role in the space station project, including its assignment to supply robotic arms for servicing spacecraft and for building the manned laboratory. Illustrated with color and black-and-white photographs.

1103. Firschein, Oscar, et al. *Artificial Intelligence for Space Station Automation*. Park Ridge, NJ: Noyes Publications, 1986. 386 p.

Includes chapter on teleoperation and robotics, discussing teleoperation and robotics requirements for satellite servicing, manufacturing in space, and orbital construction, with an extensive bibliography on this subject.

1104. Fisher, Arthur. "Robot Space Tug." *Popular Science* 229 (November 1986), 12.

Describes plans to develop the first robotic vehicle, the Orbital Maneuvering Vehicle (OMV), an unmanned space tug for pulling satellites and other orbiting objects from place to place. Illustrated with drawing.

1105. Fisher, Arthur. "Telepresence in Space." *Popular Science* 224 (May 1984), 12, 16.

Describes plans for a telepresence activator, a robotic device for space jobs, controlled by an earthbound operator, under study at the National Aeronautics and Space Administration's Marshall Space Flight Center, Huntsville, Alabama. Illustrated with drawing.

1106. Foley, Theresa M. "Commercialization Policy Threatens Space Station's Robotic Contract." *Aviation Week & Space Technology* 130 (February 6, 1989), 30-31.

Discusses plans for a Flight Telerobic Servicer, a robotic servicing unit for use in space station assembly. Includes color drawings.

1107. Foley, Theresa M. "Goddard Will Lead Space Station Automation and Robotics Work." *Aviation Week/Space Technology* 125 (September 1, 1986), 215, 217, 220.

Reports on plans at the National Aeronautics and Space Administration's Goddard Space Flight Center, Greenbelt, Maryland, to develop a station flight telerobotic server which, coupled with a robotic arm, will be used in assembling a space station and for doing other in-space tasks. Includes color and black-and-white drawings and color photograph.

1108. Foley, Theresa M. "SDIO Plans Robotic System to Service Weapons in Space." *Aviation Week/Space Technology* 128 (February 22, 1988), 45-46.

Reports on plans by the Strategic Defense Initiative Organization to determine the usefulness and cost benefits of a robotic system for refueling, repair, and other servicing of military satellites. Includes drawing and color photograph.

1109. Freitas, Robert, Jr. "Roboclone." *Omni* 5 (July 1983), 44-47, 117-118.

Describes proposed plans by the National Aeronautics and Space Administration to put a self-replicating factory on the moon made up of robots that can replicate themselves, and discusses how self-replicating robots could be made and problems the robots might face. Illustrated with paintings by Marshall Arisman.

1110. Gatland, Kenneth. *Robot Explorers*. Illustrated by John W. Wood, et al. London: Blandford Press, Ltd., 1972. 251 p.

Covers United States and Soviet space robots such as Pioneer, Mariner, Surveyor, and Luna series. Includes drawings in color of each space craft. Although dated, provides good background information on early space robots.

1111. Gibson, Bob D. "A Robot-Built Lunar Observatory." *Astronomy* 17 (July 1989), 14.

Describes robot rover that could construct a radio telescope on the moon, being researched by Battelle Columbus Laboratories, Columbus, Ohio. Illustrated with drawing.

1112. Goldberg, Jeff. "Desperately Seeking Cyborg Flight." *Omni* 11 (August 1989), 12, 64.

Reports on Honeybee Robotics, New York City, and their work on the end effector for a space station builder robot called the Flight Telerobotic Servicer. Illustrated with drawing.

1113. Golden, Frederic, and Hannifin, Jerry. "Putting an Arm on Space." *Time* 118 (November 9, 1981), 82, 86.
Reports on planned uses of a robotic arm on the space shuttle *Columbia*'s flight in November 1981. Illustrated with color drawing.

1114. Gregory, William H. "Canada Adapting Shuttle Remote Arm to Space Station Service Facility." *Aviation Week/Space Technology* 124 (March 3, 1986), 73-76.
Reports on robot arm (shuttle remote manipulator system) being built in Canada. Includes drawing.

1115. Gregory, William H. "Canada Redesigns Proposed Station Servicing Facility." *Aviation Week/Space Technology* 124 (March 17, 1986), 61-62.
Reports on proposed space station servicing facility which includes manipulator (robotic) arms. Includes drawing.

1116. Jones, Pat. "First the Robots." *Space World* X-5-281 (May 1987), 29.
Reports on Soviet Union's Mars exploration program, including the use of rovers and robot "moles" for soil testing.

1117. Kofler, Linda L. "Reach Out and Touch, Canadian Style." *Space World* W-9-273 (September 1986), 21-23.
Reports on remote manipulator arms being developed for the space station by the National Research Council of Canada. Illustrated with photograph and drawing.

1118. Nobbe, Thomas A. "Shuttle Arm to Gain Eye, Brain and Muscle." *Machine Design* 58 (April 10, 1986), 22-28.
Describes the remote manipulator system used on the space shuttle, what it is, how it works, what it does, how the vision system works, and how the next generation of manipulator systems will be different. Illustrated with color photographs and drawings.

1119. Nordwall, Bruce D. "Martin Marietta Robotic Device Will Aid in Space Station Assembly." *Aviation Week/Space Technology* 130 (May 15, 1989), 53.

Describes the Flight Telerobotic Servicer, a robot for use in assembling a space station. Illustrated with photograph.

1120. Oberg, Alcestis R. "Space Robots." *Omni* 6 (August 1984), 26.

Reports on studies on the use of robots controlled by telepresence, a remote-control technique, in space. Illustrated with photograph.

1121. Sheridan, Thomas B.; Krusser, Dana S.; and Deutsch, Stanley, eds. *Human Factors in Automated and Robotic Space Systems: Proceedings of a Symposium (Washington, D. C., January 29-30, 1987)*. Washington: National Research Council, Committee on Human Factors, Commission on Behavioral and Social Sciences and Education, 1987. 465 p.

Includes section on telepresence and telerobotics.

1122. Shifrin, Carole A. "Soviets Consider Possible Mars Rover, Sample Return Missions." *Aviation Week/Space Technology* 126 (March 23, 1987), 26-27.

Includes description of plans by Soviet Union to use robot "moles" to penetrate the soil on Mars to obtain samples.

1123. Smith, Bruce A. "Future Soviet Space Exploration to Focus on Mars, Asteroids." *Aviation Week/Space Technology* (June 22, 1987), 81, 84-85.

Includes discussion on how the Mars rover, which has robot arms, would operate on Mars. Illustrated with photographs, drawings, and a color drawing of the Mars rover in action, collecting sample material.

1124. Smith, Bruce A. "U. S./Soviet Teleconference Highlights Mars Exploration Program Objectives." *Aviation Week/Space Technology* 127 (August 3, 1987), 61, 63.

Includes discussion on ways robotic vehicles could be used and controlled on Mars. Box shows color drawing of Mars rover in proposed aeroshell structure to protect it before landing.

1125. Spangenburg, Ray, and Moser, Diane. "Why Can't a Robot be More Like a Grad Student?" *Space World* Y-2-290 (February 1988), 7-10.
Discusses research on artificial intelligence for uses in space exploration, especially with robots, being conducted at the National Aeronautics and Space Administration's Ames Research Center, Moffett Field, California. Illustrated with photographs and drawings.

1126. Stover, Dawn. "What's New...Seven-Legged Rover." *Popular Science* 234 (June 1989), 37.
Reports on a "walking beam" robot design being studied for use on Mars. Illustrated with color drawing.

1127. Sutphin, Susan. "Lunar Ramblin' Wreck." *Space World* Y-4-292 (April 1988), 22.
Reports on lunar robot developed by faculty and students at Georgia Institute of Technology, Atlanta. Illustrated with photograph.

1128. Weaver, Lelland A. C. "Factories in Space; The Role of Robots." *The Futurist* 11 (May/June 1987), 29-34.
Discusses automated tooling and systems, already available, that could be used in automated factories in space. Includes description of the Remotely Operated Service Arm (ROSA), a robot designed for nuclear power plant operations that could be used in space factories. Includes color photographs and color drawing. Cover of this issue shows a space factory and a space robot planned for use on the NASA space station, in color.

Agricultural Robots

1129. "Robot Harvesting vs. Manual Labor." *USA Today* 116 (June 1988), 13.

Reports on the use of robots for fruit harvesting. Illustrated with photograph.

1130. "Robots Invade the Farm." *Science Digest* 92 (May 1984), 32.

Reports on various types of robots being designed and developed for use in agriculture. Includes photograph of robot shearing a sheep.

1131. Aldersey-Williams, Hugh; Hunter, David; and Smith, Emily T. "Robots Head for the Farm; 'Agrimation' Will Cut Costs Sharply--But It Won't Start Replacing Farmworkers for a Decade." *Business Week* no. 2963 (September 8, 1986), 66-67.

Describes developments in designing robots to be used in "agrimation" - automated farming. Uses discussed include robotic orange pickers, apple pickers, and other fruit pickers. Includes photograph.

1132. Allen, Danita. "Robots: Farm Workers of the Future." *Successful Farming* 82 (December 1984), 10D.

Reports on ways robots are being used in agriculture.

1133. American Society of Agricultural Engineers. *Robotics and Intelligent Machines in Agriculture*. St. Joseph, MI: American Society of Agricultural Engineers, 1984. 155 p.

A collection of papers on the use of robots in agriculture from the First International Conference on Robotics and Intelligent Machines in Agriculture, held in Tampa, Florida, in 1983. Illustrated with photographs and drawings.

1134. Beck, B. "Machine." *Blaire & Ketchem's Country Journal* 12 (January 1985), 22.
Describes robot sheep-shearer used in Australia.

1135. Nobbe, George. "Fruit Robot." *Omni* 9 (February 1987), 113.
Reports on robot orange picker being developed by David Shoup and Roy Harrell at the University of Florida in Gainesville.

1136. Rodale, Robert. "People Can Outgarden Robots." *Organic Gardening* 30 (October 1983), 22-24.
Discusses implications to workers of the use of robots in the garden as well as in other areas.

1137. Rodale, Robert. "Will Robots Learn to Garden?" *Organic Gardening* 30 (July 1983), 22-24.
Discusses implications of the use of robots for gardening and possible economic consequences of robot use. Illustrated with drawing.

1138. Spahr, Sid. "Coming: New Dairy Ideas from Europe and Israel." *Successful Farming* 85 (October 1987), 50.
Includes descriptions of robot milkers being developed in Holland and Germany. Illustrated with color photograph.

1139. Tigner, Brooks. "Moo! Those Hands Are Cold." *Business Week* no. 2963 (September 8, 1986), 67.
Reports on "cowbots," robots designed to feed dairy cows, hook them up to milking machines, and clean the equipment when milking is finished. The robots are being developed by Vicon, a Netherlands-based farm equipment company.

1140. Weber, David M. "Robots Find Jobs Down on the Farm." *Electronics Week* 57 (August 20, 1984), 30-31.
Discusses projects to develop robots that milk cows, pick fruit, and do other farm-related work.

1141. Zimmerman, Mark D. "Far-Out Farming." *Machine Design* 57 (April 11, 1985), 24-25, 28, 30.
Includes descriptions of how robots can be used in agriculture. Illustrated with photographs and drawings.

Film, Television, and Theater Robots

1142. "Not Quite Human II." *The Disney Channel Magazine* (September/October 1989), 16, 28.
Describes film *Not Quite Human II*, sequel to *Not Quite Human*, about android teenager and his adventures with humans. Illustrated with color photographs.

1143. "Robot Redford." *Discover* 9 (June 1988), 8, 10.
Describes how the robots who "played" Number 5 (called Johnny Five) were "taught" for film *Short Circuit 2* by an actor (roboteer) wearing a telemetry suit that sensed motions of the actor's body and translated them into signals sent to the robot. Includes color photograph.

1144. Bierman, James H. "The Walt Disney Robot Dramas." *The Yale Review* 66 (December 1976), 223-236.
Discusses play *The General Electric Carousel of Progress* performed entirely by robots, at the 1964-1965 New York World's Fair, Disneyland, and Walt Disney World, and the development of audio-animatronics, the use of robots in films, plays, and entertainment, considering especially their use by Disney.

1145. Gabriel, Judy. "A Movie Robot Is Born." *American Cinematographer* 63 (February 1982), 128-129, 184-190.
Describes development, construction, and use of Phil, a robot that played the lead role in Universal movie *Heartbeeps*. Illustrated with photographs and drawings.

1146. Horgan, John. "Robots Rampant." *Scientific American* 259 (August 1988), 28.

Describes performances of the robots of Survival Research Laboratories, San Francisco, California. Illustrated with photograph.

1147. Lowell, Allen D. "Robots on Rampage in Shopping Mall!" *American Cinematographer* 67 (May 1986), 77-80, 82.
Describes development, construction, and use of robots in Trinity Pictures film *Killbots*.

1148. Macek, Carl. *Robotech Art 3. The Sentinels*. Norfolk, VA: The Donning Company, 1988. 186 p.
Tells the story behind the making of the sentinels in the animated "Robotech" series and examines each character and the story plot of the television series and motion picture. Illustrated with color drawings. See numbers 1149, 1151, and 1152 for more information on how this series was produced, numbers 643 and 644 for information on an 18-title fiction series based on this series, and number 650 for a game based on the series.

1149. Macek, Carl, and Baron, Mike. *Robotech the Graphic Novel: Genesis: Robotech*. Illustrated by Neil D. Vokes, Ken Steacy, Bob Pinaha, and Tom Vincent. Norristown, PA: Comico the Comic Company, 1986. 48 p.
Using comic book format, provides the background history of characters and robots in the "Robotech" television series. See numbers 1148, 1151, and 1152 for more information on how this series was produced, numbers 643 and 644 for information on an 18-title fiction series based on this series, and number 650 for a game based on the series.

1150. Magid, Ron. "Robotrix for *Short Circuit 2*." *American Cinematographer* 69 (September 1988), 78-80, 82, 84, 86.
Describes development and use of robot Johnny Five in film *Short Circuit 2*. Illustrated with photographs.

1151. Reynolds, Kay, and Carlton, Ardith. *Robotech Art 1*. Norfolk/Virginia Beach, VA: The Donning Company, 1986. 254 p.

Provides summaries of 85 episodes of the automated "Robotech" television series, information on characters and equipment, historic background on the "Robotech" idea, and background information on persons producing the series. Illustrated with color drawings from the series and black-and-white drawings and photographs. See numbers 1148, 1149, and 1152 for more information on how the series was produced, numbers 643 and 644 for information on an 18-title fiction series based on the series, and number 650 for a game based on the series.

1152. Reynolds, Kay, ed. *Robotech Art 2*. Norfolk/Virginia Beach, VA: The Donning Company, 1987. 131 p.

Covers the characters in the animated "Robotech" television series and includes a history of robot fantasy stories, comics, and animation in Japan. Illustrated with photographs of artists and color drawings from the series. See numbers 1148, 1149, and 1151 for more information on how this series was produced, numbers 643 and 644 for information on an 18-title fiction series based on this series, and number 650 for a game based on the series.

1153. Schaefer, Stephen. "Robots, the Men." *Film Comment* 22 (July/August 1986), 6-7.

Discusses the movie robot and current robot movies. Illustrated with photographs.

1154. van Gelder, Lawrence. "Syd Mead's No. 5, Giving a Robot a Soul." *New York Times* 135 (May 9, 1986), C10.

Reports on Syd Mead, the man who designed Number 5 (Johnny Five), the robot in the film *Short Circuit*. Illustrated with photograph.

1155. Weber, Bruce. "Method Acting." *New York Times Magazine* no. 5 (December 20, 1987), 86.

Describes ways telemetry suit was used to make robot Number 5's actions more humanlike in film *Short Circuit 2*. Includes color photograph.

1156. Weisang, Myriam. "Masters of the Universe." *Theatre Crafts* 21 (January 1987), 28-29, 51-53.
Describes the machine performances of robots, staged by Survival Research Laboratories, San Francisco, California. Illustrated with color and black-and-white photographs.

1157. Zorpette, Glenn. "Art Among the Ruins." *Art News* 87 (September 1988), 16.
Describes New York show put on by robots of Survival Research Laboratories, San Francisco, California. Illustrated with color photograph.

Fine Arts Robots: Art and Music

1158. "Robots in Review." *Americana* 12 (March/April 1984), 112.
Provides color photographs of four robots from an exhibit of 28 at the American Craft Museum II in New York City, in the first major museum show devoted to a historical view of robots. Guest curator was Robert Malone.

1159. "Will the Real Andy Warhol Please Stand up and Say Something?" *Science Digest* 92 (April 1984), 85.
Reports on robot likeness of Andy Warhol being built by show producer Lewis Allen to star in one-robot show.

1160. Amato, Ivan. "The Finishing Touch." *Science News* 135 (February 18, 1989), 108-09.
Describes the development of a robot to do final finish rubbing on a Steinway piano. Illustrated with photograph.

1161. Haddock, Vickie. "Spare Parts." *American Craft* 44 (April/May 1984), 14-17.
Describes work of robot-builder artist Clayton Bailey in California. Some of his robots are wired for action. Includes color and black-and-white photographs.

1162. Hagans, Orville R. "The Defender of Time." *Hobbies* 87 (January 1983), 72-73.
Describes robot clock called "The Defender of Time" in the "Clock District" in Paris, created by Jacques Monestier, French engineer, architect, and robot enthusiast. Includes photographs.

1163. Hawkins, William J. "Electronics Newsfront." *Popular Science* 230 (June 1, 1987), 22-23.
Includes description of Panasonic's robotic portrait maker that can "see" its subject and use the data to brush-draw a portrait. Includes color photograph.

1164. Holt, Steve. "A Fantasy Portfolio: Robots." *Industrial Design* 32 (September/October 1985), 54-59.
A gallery of photographs and drawings of fantasy personal robots, submitted by invited designers.

1165. Lecht, Charles P. "Robot Pipes Technology into the Beaux Arts." *Computerworld* 19 (April 29, 1985), 37.
Describes Wasubot, Japanese organ-playing robot.

1166. Mann, Marjorie B. "Sculpture." *Omni* 5 (April 1983), 132.
Describes robot sculptures created by Catherine Field of New York City. Illustrated with photograph.

1167. Peterson, Ivars. "The Quality of Violin Strings." *Science News* 128 (November 16, 1985), 313.
Describes robotic violin player, a "bowing machine" which is used to simulate a person playing the violin in order to test the strings.

1168. Reilly, Patti. "Tim Pace's Band Doesn't Touch Drugs or Groupies, but It Wouldn't Go Far without Juice." *People Weekly* 14 (December 8, 1980), 94.
Reports on the Nuclear Diode Robot Band, a four-robot band built by Tim Pace of Falls Church, Virginia.

Food Handling and Restaurant Robots

1169. "Burgers with Cheese, Relish, Fries ... and Robot." *New Scientist* 116 (October 8, 1987), 31.

Describes robot being used to assemble hamburger sandwiches at the University of Wisconsin - Stout in Menomonie. Illustrated with drawing.

1170. "Confucius Say: Robot Waiters Reap Fortune, Cookie (But What's That Transistor Doing in My Soup?)." *People Weekly* 19 (June 13, 1983), 55.

Reports on robot waiters at the Two Panda Deli, a Chinese fast food restaurant in Pasadena, California. Two Japanese robots deliver food, play music, speak, and react to specific voice commands. Similar robots have been used in Japan since the early 1970's, according to the owners. Includes photograph.

1171. "Nolan Bushnell's New Act: Singing Robots." *Fortune* 104 (July 27, 1981), 44.

Describes use of pneumatically powered robots for entertainment at Pizza Time Theatres, fast food-video arcades. Illustrated with photograph.

1172. "Roboburger." *Discover* 9 (May 1988), 6.

Discusses the use of robots in fast food restaurants and a restaurant robot project directed by Tom Phillips at the University of Wisconsin in Menomonie. Includes color drawing.

1173. "'Robots' Stack up Just Fine as Man's Cold Weather Friend." *Quick Frozen Foods* 47 (February 1983), 42-43.

Describes automatic stacker cranes used in a frozen food storage center. Illustrated with photographs.

1174. Bonnett, Jim. "Counterploy in the High Street." *Management Today* (September 1986), 127-128.

Describes supermarket in Japan that uses robots as butchers, egg and produce sizers and testers, and stackers, and in other capacities.

1175. Curtis, John W. "Robotics and Artificial Intelligence for the Food Industry." *Food Technology* 41 (December 1987), 62-64.

Examines possible applications and pitfalls of robotics and artificial intelligence in the food industry.

1176. Inman, Virginia. "The Silicon Saloonkeeper." *Inc.* 6 (September 1984), 23-24.

Describes robot bartender developed by Scarab Robotics Corporation, San Francisco, California. Illustrated with photographs.

1177. Kanner, Bernice. "Automation House." *New York* 18 (April 1, 1985), 34.

Describes plans by Hughes International for restaurant in New York City staffed entirely by robots. Illustrated with drawing.

1178. Kinkead, Gwen. "High Profits from a Weird Pizza Combination." *Fortune* 106 (July 26, 1982), 62-66.

A business update on two fast food restaurants, Pizza Time Theatre and ShowBiz Pizza Place, that use robots as part of their entertainment while customers are waiting for their pizzas. Illustrated.

1179. Lingle, Rick. "Robotics: Thought for Food." *Prepared Foods* 154 (November 1985), 61-62, 64.

Reports on robot use in the food industry, with inserts on use in food analysis laboratories, in meat processing, and for palleting (stacking and transporting cartons and cases). Includes photographs and graph.

1180. Mans, Jack. "Robotics: Blueprint for the Future." *Processed Prepared Foods* 152 (November 1983), 68-70, 72.

Discusses ways robots are used in food preparation plants. Illustrated with photographs and drawing.

1181. Nobbe, George. "Chocolate Robot." *Omni* 9 (June 1987), 42.
Reports on chocolate-decorating robot developed by A. J. Cronshaw and others at PA Technology in Cambridge, England.

1182. Phillips, Thomas R. "Robotics Revolution." *Restaurant Business* 86 (May 20, 1987), 169-171.
Reports on uses of robots in the food service industry. Illustrated with drawings.

1183. Riell, Howard. "Is There a Robot in Your Future?" *Nation's Restaurant News* 20 (January 20, 1986), 40.
Discusses the use of robots in restaurants.

1184. van Warner, Rick. "Robotics: An Alternate Labor Source." *Nation's Restaurant News* 21 (August 17, 1987), 18.
Reports on robot applications in the fast food industry.

1185. van Warner, Rick. "Robots: A Possible Answer to Industry's Manpower Crisis." *Nation's Restaurant News* 21 (April 20, 1987), 15.
Reports on uses of robots in the food service industry.

Handicapped Users and Uses

Nonfiction

1186. "A Helping Hand for the Blind." *Design News* 44 (August 1, 1988), 48.
Reports on Dexter, a robotic hand that finger spells messages sent from a computer keyboard, being developed at Stanford University, Stanford, California. Illustrated with photograph.

1187. "Aiding the Disabled and the Elderly." *USA Today* 113 (June 1985), 14.

Reports on ways robots can be used to help in hospitals and with the disabled and elderly according to Professor Larry Leifer, head of a team at the Veterans Administration Medical Center at Palo Alto, California, and Paul Haber, regional director of programs for the elderly there. Includes photograph.

1188. "New Hope for Disabled." *USA Today* 111 (June 1983), 10-11.

Reports on work on rehabilitative robotics being done by a team directed by Professor Larry Leifer at the Veterans Administration Rehabilitation Research and Development Center in Palo Alto, California. Includes photograph.

1189. "Robot Lets the Fingers Do the Talking." *New Scientist* 118 (June 16, 1988), 47.

Reports on Dexter, a "talking" glove and mechanical hand, designed to help handicapped people communicate more effectively, developed at Stanford University, Stanford, California. Illustrated with photograph.

1190. "Robot Serves as Limbs for Quadriplegic." *Design News* 43 (November 23, 1987), 52-53.

Describes robot-assist workstation for handicapped workers, designed and built by engineers at Georgia Institute of Technology, Atlanta, and used by Phil Payne at the Georgia Department of Human Resources. Illustrated with photograph.

1191. "Robotic Arms Aid Handicapped Kids." *USA Today* 117 (February 1989), 6-7.

Reports on uses of robotic arms with children with severe physical disabilities, being studied at Ohio State University, Columbus. Illustrated with photograph.

1192. Bak, David J. "Automated Care Station Frees Severely Disabled." *Design News* 42 (October 20, 1986), 118-119.

Describes computerized bedside system which includes voice-activated MiniMover 5 robot arm, developed at the Office of Interdisciplinary Programs, Georgia Institute of Technology, Atlanta. Illustrated with photographs.

1193. Chase, Marilyn. "Voice-Activated Robot to Aid Quadriplegics Is Research Aim." *The Wall Street Journal* 208 (August 29, 1986), 19.
Describes robots being developed to aid quadriplegics.

1194. Engelhardt, K. G.; Edwards, Roger; Sample, Whitney; and Sandrof, Michael. "The Robotic Vocational Workstation." In *1988 Annual Research Review*, pp. 66-79. Edited by Nicole S. Vecchi. Pittsburgh, PA: Carnegie Mellon University, The Robotics Institute, 1989, 111 p.
Describes research on workstation using a robot manipulator for disabled individuals, at Carnegie Mellon University. Illustrated with photographs.

1195. Foulds, Richard E. *Interactive Robotic Aids - One Option for Independent Living: An International Perspective. Monograph Number 37.* New York: World Rehabilitation Fund, Inc., 1986. 71 p. ERIC document ED 285 374.
A collection of papers on the role of robotics in rehabilitation. Includes "A Potential Application in Early Education and a Possible Role for a Vision System in a Workstation Based Robotic Aid for Physically Disabled Persons," by W. Harwin, et al.

1196. Hales, Dianne. "To Build a Better Robot." *Family Health* 13 (March 1981), 24.
Reports on robots being developed to help the disabled at the Veterans Administration Rehabilitative Engineering and Development Center in Palo Alto, California. Includes photograph.

1197. Hoban, Phoebe. "Robot Nurses." *Omni* 6 (December 1983), 28, 188.

Describes robot used to help the handicapped, developed at the Rehabilitation Research and Development Center, Palo Alto, California, and reports on other research projects on robots for the handicapped. Illustrated with photograph.

1198. Kimbler, D. L. "Robots and Special Education: The Robot as Extension of Self." *Peabody Journal of Education* 62 (Fall 1984), 67-76.

Discusses use of robots with handicapped students and performance characteristics and capabilities robot should have to be effective.

1199. Kinnaman, Daniel E. "Robot Guide Dog: Will It Replace Man's Best Friend?" *Classroom Computer Learning* 10 (October 1989), 32G.

Describes a mobile robot being designed at North Carolina State University, Raleigh, to assist the visually impaired in a known environment. Illustrated with drawing.

1200. Leifer, Larry. "Rehabilitative Robots." *Robotics Age* 3 (May/June 1981). Reprinted in *Robotics Age: In the Beginning*, pp. 227-238. Edited by Carl Helmers. Rochelle Park, NJ: Hayden Book Company, Inc., 1983. 241 p.

Examines the objectives and evolution of rehabilitative robotics, describes work being done in this area at the Rehabilitative Engineering Research and Development Center at the Palo Alto Veteran's Administration hospital in Palo Alto, California, and defines specifications for an interactive robotic aid for the severely disabled. Four areas being studied for possible use of robotic technology include: (1) in daily living activities (for example, preparing food, eating, and personal hygiene); (2) in personal clerical tasks (operating a telephone, calculator, or computer); (3) in vocational tasks (computer programming or secretarial work); (4) in recreation (for example, controlling electronic games, playing chess, or painting). Illustrated with photographs and drawings.

1201. Mishara, Eric. "Robot for the Handicapped." *Omni* 6 (January 1984), 39.
Reports on robot used with quadriplegics at Veterans Administration Medical Center, Palo Alto, California. Illustrated with photograph.

1203. Moore, Gwendolyn B.; Yin, Robert K.; and Lahm, Elizabeth A. "Robotics, Artificial Intelligence, Computer Simulation: Future Applications in Special Education." *Technological Horizons in Education Journal* 14 (August 1986), 74-76.
Reports on results of a study to determine how robotics, artificial intelligence, and computer simulation can be used by handicapped students, based on current applications of these technologies. Study concludes that technical barriers to robotics applications primarily involve the development of adequate auxiliary control systems and flexibility for use in everyday settings.

1204. Moore, Gwendolyn B.; Yin, Robert K.; and Lahm, Elizabeth A. *Robotics, Artificial Intelligence, Computer Simulation: Future Applications in Special Education*. Washington, DC: COSMOS Corporation, 1985. 163 p. ERIC document ED 261 496.
Includes discussion on current applications of robots and future uses of robotics in special education.

1205. Smith, Emily T. "These Robots Give the Disabled a Hand." *Business Week* no. 3057 (June 20, 1988), 103.
Reports on two robots designed to help the handicapped - a robotic arm designed to help stroke and accident victims perform rehabilitative exercises and a voice-controlled workstation for the severely disabled - both available from Prab Command, Incorporated, of Kalamazoo, Michigan.

1206. Yin, Robert K., and Moore, Gwendolyn B. "The Use of Advanced Technologies in Special Education: Prospects from Robotics, Artificial Intelligence, and Computer Simulation." *Journal of Learning Disabilities* 20 (January 1987), 60-63.

Describes study to develop a general methodology for continuing to identify potential uses of new technologies for special education, and to apply this methodology to three advanced technologies.

Fiction

1207. Caidin, Martin. *Cyborg*. New York: Arbor House, 1972. 282 p.

Relates the story and adventures of a military test pilot whose legs, left arm, and one eye are replaced with artificial parts having superhuman strength and unusual capabilities, after he is severely injured in the crash of a plane he is testing. Television series, "The Six-Million Dollar Man," was based upon this novel.

1208. Caidin, Martin. *Cyborg IV*. New York: Arbor House, 1975. 205 p.

Continues the story and adventures of a military test pilot with artificial parts with superhuman strength and unusual capabilities.

Health Care and Medical Robots

1209. "At a California Hospital, the Surgeon with the Steadiest Hand Is a Robot Named Ole." *People Weekly* 23 (May 6, 1985), 97.

Reports on the world's first neurosurgical robot, Ole, developed by a team led by Yik San Kwoh at Memorial Medical Center in Long Beach, California.

1210. "Brain Surgeon's Robot Helper." *Popular Mechanics* 165 (June 1988), 17.

Reports on robotic system that constantly repositions a brain surgeon's microscope, developed at Carnegie Mellon

University, Pittsburgh, Pennsylvania. Illustrated with color photograph.

1211. "Coming: Robot Servants?" *Design News* 44 (March 21, 1988), 50-51.

Describes Roscoe, a mobile robot designed to haul trays in hospitals and clinics, developed by Transitions Research Corporation, Danbury, Connecticut (founded by Joseph F. Engelberger), and other mobile robots. Illustrated with drawing.

1212. "Doctors/Robots Team up to Implant Hips." *Design News* 44 (August 1, 1988), 34, 37.

Reports on research on how a robot could help in hip replacement. Illustrated with color drawing.

1213. "Finding the Perfect Fit for Hip Replacement." *Insight* 4 (May 30, 1988), 52.

Describes robotic device that allows a more precise fitting of hip replacements than procedures based on measurements taken by a surgeon, allowing implants to be held in place without cement. Includes drawing.

1214. "Robots Could Slice into Manpower Needs." *Hospitals, Journal of American Hospital Association* 58 (April 1, 1984), 54.

Discusses ways robots could be used in hospitals. Illustrated with photograph.

1215. "Robots Detect Breast Cancer." *The Futurist* 23 (March/April 1989), 5.

Reports on robot that can detect cancerous lumps, developed at Waseda University, Japan.

1216. "Will Your Doctor Be a Robot?" *USA Today* 117 (February 1989), 9.

Reports on ways robots are used in the health care field.

1217. Blakeslee, Sandra. "Robot Assists in Surgery." *Reader's Digest* 127 (November 1985), 29-30.
Reports on computerized robotic arm developed by Yik San Kwoh and colleagues at Memorial Medical Center in Long Beach California, for use in brain surgery.

1218. Boissoneau, Robert, and Anderson, David. "Robots May Give a Hand in Health Care." *Modern Health Care* 13 (June 1983), 89.
Discusses ways robots could be used in hospitals and nursing homes.

1219. Emert, Victor. "Foobie and the Million Dollar Machine: High Tech Comes to Drug Education." *Adolescent Counselor* April/May 1988, 26-28.
Describes the use of a robot named Foobie D. Robot for drug education in schools, which is available from David GoWell and Kent Davis, 2369 Phillips Road, Mount Holly, NJ 08060, telephone 609-261-2162. Illustrated with photographs.

1220. Freifeld, Karen. "Right Hand Robot." *Health* 17 (September 1985), 19.
Reports on the use of a robot arm, named Ole, as an aid to brain surgery at Memorial Medical Center in Long Beach, California. Procedure is called "robot-aided surgery."

1221. Freiherr, Greg. "Robots Come to the OR." *Science 86* 7 (May 1986), 66.
Reports on use of robot arm developed by Yik San Kwoh at the Memorial Medical Center, Long Beach, California, for use in neurosurgery. Illustrated with photograph.

1222. Maleskey, Gale. "Battery-Powdered Candy Striper." *Prevention* 40 (June 1988), 12.
Reports on Roscoe, a robot designed to bring food trays to nurses' stations and take specimens to the laboratory at

Danbury Hospital, Danbury, Connecticut. Illustrated with color drawing.

1223. McLean, Valnere P. "Using Robots in a Group with the Neuropsychologically Impaired: Achieving a Sense of Control over the Environment." *Occupational Therapy in Mental Health* 8 (1988), 155-164.

Reports on a study which used robots in therapy.

1224. Nicholson, Tom; Willenson, Kim; and Doi, Ayako. "Robots: Japan Takes the Lead." *Newsweek* 98 (September 21, 1981), 92.

Reports on various types of robots being developed in Japan, especially for the medical field. Includes photographs.

1225. Smith, Emily T. "Robo-Doc Is Ready to Help in the Operating Room." *Business Week* no. 3043 (March 21, 1988), 15.

Reports on a robot to assist surgeons in hip operations, being developed by International Business Machines (IBM) and the University of California at Davis. Illustrated with color drawing.

1226. Stedman, Nancy. "The Robot's Touch." *Health* 15 (November 1983), 12.

Reports on Clancy, a hospital robot used to deliver laboratory samples at Yale-New Haven Hospital in Connecticut.

1227. Tierney, John. "Yik San Kwoh Built a Better Brain Surgeon." *Esquire* 104 (December 1985), 218-219, 221.

Tells the story of how Ole, the robot used in brain surgery, was developed and tested by Yik San Kwoh, and how it was first used. Includes color photograph.

1228. Wilkinson, Richard, and Rudeseal, Pat B. "Robots Move Supplies for Less Money and Labor." *Hospitals* (December 5, 1986), 70.

Describes automatic guided vehicle used at the University of Michigan Hospitals, Ann Arbor, Michigan.

1229. Wolkomir, Richard. "Long-Distance Operators." *Omni* 10 (May 1988), 28, 135.

Discusses the Virtual Environmental Work Station project which aims to design a workstation to perform telesurgery, using robot telesurgeons, and do other medical work by remote operation, at the National Aeronautics and Space Administration's Ames Research Center, Moffett Field, California. Illustrated with color photograph.

1230. Zweig, C. "Work in Progress." *Esquire* 102 (November 1984), 253.

Reports on and describes a robot to be used for performing brain surgery, a computerized arm with an imaging device to assist doctors in neurosurgery at Memorial Medical Center in Long Beach, California. Advantages include pinpoint precision and stability. Use of the robot shortens the hospital stay.

Laboratory Robots

1231. "Robot Tackles Chemistry's Dirty Side." *USA Today* 115 (June 1987), 8-9.

Describes development work on robot arm for use with chemistry experiments and chemicals in laboratories, by Phil Fuchs and Gary Kramer at Purdue University, West Lafayette, Indiana. Includes photograph.

1232. "Robots Find Use in Chemical Synthesis Labs." *Chemical & Engineering News* 62 (September 10, 1984), 31.

Reports on use of robot to aid optimization of chemical reaction conditions.

1233. "Send in the Robots." *Beverage World* 104 (December 1985), 34.

Describes the Zymate robot used by Coca-Cola USA's central laboratory to conduct quality assurance tests by analyzing samples of Coca-Cola. Illustrated with photograph.

1234. Barnes, Deborah M. "Robot War Weeds Out Competitors." *Science* 239 (March 18, 1988), 1376.
Describes robots used for laboratory work. Illustrated with photograph.

1235. Bierman, Howard. "Robots Invade Chemistry Labs." *Electronics Week* 58 (March 11, 1985), 40.
Describes robots developed for testing, analysis, and preparation of samples in the laboratory, on display at the Pittsburgh Conference on Analytical Chemistry and Applied Spectroscopy held in February 1985. Includes photographs.

1236. Hurst, William Jeffrey, and Mortimer, James W. *Laboratory Robotics*. New York: VCH Publishers, Inc., 1987. 129 p.
Covers need for robotics in the laboratory, automation through laboratory unit operations, justification of a laboratory robot, applications, getting started, programming, and future of laboratory robots. Illustrated with photographs and drawings.

1237. Kerr, Richard A. "Sample Preparation by a One-Armed Robot." *Science* 216 (April 9, 1982), 164.
Reports on a sample preparation robot manufactured by Zymark Corporation of Hopkinton, Massachusetts, displayed at the 1982 Pittsburgh Conference on Analytical Chemistry and Applied Spectroscopy held in Atlantic City.

1238. Kramer, Gary W., and Fuchs, Philip L. "Automation in Organic Synthesis." *Byte* 11 (January 1986), 263-264, 266, 268, 270, 272-274, 276, 278, 280, 283-284.
Includes description of how a robot is used to handle sample preparation and other tasks in an organic chemistry laboratory at Purdue University, West Lafayette, Indiana. Illustrated with photograph and drawings.

1239. Robinson, Arthur L. "Robots Automate Sample Preparation." *Science* 227 (March 29, 1985), 1565-1566, 1568-1569.

Describes robots developed for preparation of samples in the laboratory, on display at the Pittsburgh Conference on Analytical Chemistry and Applied Spectroscopy held in February 1985, in New Orleans, Louisiana.

1240. Strimaitis, Janet R. "Robots in the Laboratory - An Overview." *Journal of Chemical Education* 67 (January 1990), A20, A22.

Discusses applications, precision and accuracy, and system integration of robots in scientific laboratories.

1241. Wolfram, Lynn E. "Lab Robot Prepares Analytical Samples." *Research & Development* 28 (July 1986), 74-77.

Describes uses of laboratory robot at Standard Oil Company of Ohio, Cleveland, Ohio. Illustrated with drawing.

Military Robots

1242. "Foot Soldiers to Get Suits of Armor?" *Machine Design* 58 (October 9, 1986), 20.

Describes PITMAN, a "mind-reading" suit for the infantry, proposed by Jeffrey Moore, Advanced Weapons Technology Group, Los Alamos National Laboratory, New Mexico.

1243. "GI Robots." *The Futurist* 17 (February 1983), 29.

Summarizes a United States Army report, *Concepts for Army Use of Robotic-Artificial Intelligence in the 21st Century*, by Dennis V. Crumley, describing some of the ways robots might be used in the military.

1244. "Marines' Remotely Operated Device Will Use Aerial Fiber-Optic Link." *Aviation Week/Space Technology* 126 (February 2, 1987), 73.

Describes aerial surveillance device to use as part of overall ground air telerobotics systems (Gaters), a two-year effort to turn telerobotics concept - remote control of a robot - into usable military hardware.

1245. "Remote-Control GIs." *Popular Mechanics* 161 (July 1984), 119.

Reports on Odex l, a six-legged, walking robot made by Odetics, Incorporated of Anaheim, California, and plans for use in the United States Army. Includes color photograph and drawing.

1246. "Robot Armored Vehicles to Navigate with Lasers." *Popular Mechanics* 165 (September 1988), 16.

Reports on remotely controlled, tank navigation system being developed by Odetics, Incorporated, Anaheim, California. Illustrated with color drawings.

1247. "Robot-Loaded Howitzer Targets Battle Stress." *Design News* 40 (February 6, 1984), 18, 20.

Describes robot being developed to load self-propelled howitzer, at Picatinny Arsenal Armament Research and Development Center, Dover, New Jersey. Illustrated with photograph.

1248. "Robotic Mine Hunter." *Popular Mechanics* 165 (September 1988), 17.

Reports on remotely operated vehicle that tracks down and destroys mines, developed by Plessey Naval Systems, Templecombe, England. Illustrated with color drawing.

1249. Aronson, Robert B. "Robots Go to War." *Machine Design* 56 (December 6, 1984), 72-79.

Discusses military uses of robots and describes research projects and robotic aircraft. Illustrated with photographs and drawings.

1250. Bak, David J. "War without Men." *Design News* 44 (September 5, 1988), 67-70.

Describes an imaginary battle which includes the use of robots and reports on military robots being developed. Illustrated with color photographs and drawings.

1251. Barnaby, Frank. *The Automated Battlefield*. New York: The Free Press, 1986. 180 p.
Includes information on functionoids, robotic land vehicles, remotely piloted vehicles, and other uses of robots for military purposes.

1252. Chamberlain, Gary. "Robot Warriors." *Design News* 43 (September 7, 1987), 116-118, 120.
Describes military robot projects such as teleoperated vehicles (TOVs), remotely operated aircraft, and autonomous ammunition loaders. Illustrated with color photographs and color drawing. Cover of this issue shows a color photograph of a TOV.

1253. Corrado, Joseph K. "Military-Industrial Cooperation Advances Robotics Applications." *Mini-Micro Systems* 16 (December 1983), 122-124, 128, 130.
Describes robots being developed by or for the United States Department of Defense. Illustrated with photographs and drawing.

1254. Cunningham, Ann Marie. "Robo-Soldiers." *Technology Review* 91 (April 1988), 11-12.
Describes the PITMAN powered exoskeletal suit, possible "thinking armor" for soldiers, designed by Jeffrey Moore at Los Alamos National Laboratory, New Mexico.

1255. Davies, Owen. "Robotic Warriors Clash in Cyberwars." *Omni* 9 (January 1987), 76-78, 80-81, 88.
Reports on military robots being used now, in research, or planned for the future, in the air, on land, and underwater. Illustrated with color photograph.

1256. Fisher, Arthur. "Robotic Fire Ant." *Popular Science* 234 (January 1989), 16.

Describes a small robotic vehicle designed to detect and demolish armored vehicles, developed at Sandia National Laboratories in Albuquerque, New Mexico. Illustrated with color photograph.

1257. Fjermedal, Grant. "The Blade Runner Blues." *Omni* 9 (January 1987), 78.
Gives various futurists' comments on military interest in creating warrior robots and possible consequences.

1258. Griffiths, Dave. "Robo-Soldiers Are Inching Toward the Front." *Business Week* no. 3077 (October 31, 1988), 156-157.
Discusses military robotics projects, unmanned aircraft, robot submarines, and robot tank and other ground weapons. Includes color drawing.

1259. Klass, Philip J. "DARPA Envisions New Generation of Machine Intelligence Technology." *Aviation Week/Space Technology* 122 (April 22, 1985), 46-47, 51, 53-54.
Discusses various artificial intelligence technology projects of the Defense Advanced Research Projects Agency, including an autonomous land vehicle and six-legged vehicle. Illustrated with photographs.

1260. Kopvillem, Peeter. "The Pentagon Funds a Walking Robot." *Maclean's* 98 (November 4, 1985), 69-71.
Reports on six-legged, 17-foot-long, computerized robot called an adaptive suspension vehicle (ASV). Developed at Ohio State University, Columbus, robot is reported to be the first fully self-contained, outdoor machine. Describes how and why it may be used in areas such as forest and tundra.

1261. Menosky, Joseph. "Killer Turtles." *Science '85* (April 1985), 76-78.
Describes research and inquiry (sometimes humorously) into the military uses of turtle robots on the battlefield. Illustrated with photograph.

1262. Schefter, Jim. "Look Ma! No Driver." *Popular Science* 227 (October 1985), 94-95, 114, 116.

Describes Autonomous Land Vehicle (ALV), an unmanned vehicle driven by a vision system and/or remote control, developed by Martin Marietta Aerospace for the Defense Advanced Research Projects Agency. Illustrated with color photograph and drawing.

1263. Shaker, Steven M., and Wise, Alan R. *War without Men: Robots on the Future Battlefield*. McLean, VA: Pergamon-Brassey's International Defense Publications, Inc., 1988. 196 p.

Covers evolution of military systems, current operational use and development of unmanned robot ground vehicles, remotely piloted vehicles, unmanned naval vessels, space-based robotics, and robots in the future battlefield. Illustrated with photographs.

Nuclear and Hazardous Environment Robots

1264. "Robot 'Kid Gloves' Wafer Processing." *Design News* 43 (May 25, 1987), 36.

Reports on robot used to fabricate radiation-hardened integrated circuits at Sandia National Laboratories' Center for Radiation-Hardened Microelectronics, Albuquerque, New Mexico, the first United States research lab to use a robot in the entire production process.

1265. "Robot Scans for Toxic Waste." *Popular Mechanics* 167 (May 1990), 18.

Reports on robot for investigating hazardous waste dumps, being developed at Carnegie Mellon University, Pittsburgh, Pennsylvania. Illustrated with color photograph.

1266. "Robot Staff for Nuclear Power Plants." *Science News* 129 (March 29, 1986), 203.

Reports on Surveyor, a remotely controlled robot for use in nuclear power plants. Robot was developed for the Electric Power Research Institute at Palo Alto, California, by Advanced Research Development Corporation in Columbia, Maryland. Also reports on a walking robot being developed by Odetics, Incorporated, Anaheim, California.

1267. "Robots Clean Chernobyl Power Plant Site." *Design News* 43 (January 19, 1987), 24.
Describes West German robots being used in damage assessment and cleanup at Chernobyl Nuclear Power Plant in Soviet Union. Illustrated with color and black-and-white photographs.

1268. "Robots Maintain Reactors - and Demolish Them." *New Scientist* 94 (May 6, 1982), 356.
Reports on robot inspection devices and demolition robots for use in nuclear plants, under development by Taylor Hitec in England. Includes photograph and drawing.

1269. "Travelling Robot to Work in Radiation-Hardened IC Lab." *Radio-Electronics* 58 (November 1987), 4.
Reports on robot used to fabricate radiation-hardened integrated circuits at Sandia National Laboratories' Center for Radiation-Hardened Microelectronics, Albuquerque, New Mexico, the first United States research lab to use a robot in the entire production process.

1270. "Troubleshooting with Robots." *USA Today* 115 (June 1987), 14.
Reports on research to develop a robot for troubleshooting use in nuclear reactors.

1271. Bagne, Paul. "After the China Syndrome." *Omni* 6 (September 1984), 56-58, 60, 62, 116, 118.
Describes a visit to Three Mile Island Nuclear Power Plant to watch cleanup operations and the types of robots used or planned for use in the cleanup. Illustrated with photographs.

1272. Crawford, Mark. "Soviets Buy Robots." *Science* 232 (June 13, 1986), 1332.

Reports on Soviet purchase, from West Germany, of three specialized robots needed to probe the Chernobyl Nuclear Power Plant after the disaster.

1273. Hooper, Judith. "Advanced Telepresence. Robots for Hazardous Occupations." *Omni* 9 (October 1986), 56.

Describes the Advanced Servomanipulator, a pair of robotic arms linked electronically to a distant human operator, being used at Oak Ridge National Laboratory, Oak Ridge, Tennessee. Illustrated with photograph.

1274. Horton, Elizabeth. "Robot Wreckers." *Science Digest* 90 (September 1982), 35.

Describes a robot developed by Taylor Hitec of Lancashire, England, for use in dismantling reactor vessels in old nuclear power plants.

1275. Larcombe, M. H. E., and Halsall, J. R. *Robotics in Nuclear Engineering*. London: Graham and Trotman, Ltd. for the Commission of the European Communities, 1984. 153 p.

Describes the use of robots in nuclear engineering, with many drawings of all types of robots and their applications in work involving radioactivity.

1276. Milne, Roger. "Breaking up Is Hard to Do." *New Scientist* 112 (December 11, 1986), 34-37.

Explains how the Windscale Nuclear Reactor in Great Britain is being taken apart with the aid of robots. Illustrated with photographs.

1277. Yeaple, Frank. "Microcomputer-Controlled Robot Inspects Nuclear Plant." *Design News* 43 (March 23, 1987), 98-100.

Describes SURBOT, surveillance robot used at Brown's Ferry Nuclear Plant in Decatur, Alabama. Illustrated with photographs and drawing.

Police, Security, and Sentry Robots

1278. "A Bomb Squad Cop on Wheels." *Popular Mechanics* 160 (July 1983), 121.
Reports on Pedsco Mark-3, a Canadian-built, remotely controlled mobile robot with robotic arm, used by police departments to pick up bombs. Includes photographs.

1279. "A Disarming Robot." *Discover* 4 (September 1983), 34.
Reports on a robot used to disarm bombs. Illustrated with photograph.

1280. "Auxiliary Robotic Educational System." *The Police Chief* (April 1987), 33.
Advertisement made up of police department testimonials for uses of ARES robots made by 21st Century Robotics, Norcross, Georgia. Includes illustration.

1281. "Close Encounters of the California Kind." *Discover* 8 (November 1987), 18.
Reports on mistaken use of armed robots against unlabeled weather balloons.

1282. "Robot from Cellblock H." *Popular Mechanics* 161 (July 1984), 119.
Reports on Denny, a prison guard robot made by Denning Mobile Robotics, Woburn, Massachusetts. Includes color photographs.

1283. Adams, Jane Meredith. "A Robot Named Denny with Pinkerton's Job Makes for Hot Stock." *New England Business* 6 (February 6, 1984), 38-40.
Reports on sentry robot built by Denning Mobile Robotics, Incorporated, Woburn, Massachusetts. Illustrated with drawing.

1284. Darrow, Barbara. "Robot Vehicle Disarms Terrorist Bombs." *Design News* 44 (January 4, 1988), 43, 45.

Describes Ferret, a robot designed to dispose of bombs, developed by Rapidex, Incorporated, Saxonville, Massachusetts.

1285. Henry, Sarah. "Mounties of Steel." *Maclean's* 92 (August 27, 1979), 47.

Describes the Remote Mobile Investigation unit, a mobile robot built by Pedsco Canada, Limited. The robot can destroy a bomb, drag a wounded person to safety, fire tear gas grenades, carry food and supplies, temporarily blind a sniper by using a mirror, shoot a rifle, shoot a chemical-containing water pistol, and negotiate with terrorists through a speaker system. Includes photograph.

1286. Lampe, David. "Robot Sentries." *Popular Science* 227 (August 1985), 20.

Reports on Denny, sentry robot made by Denning Mobile Robotics, Woburn, Massachusetts, with Carnegie Mellon University's Robotics Institute in Pittsburgh, Pennsylvania. Includes photograph.

1287. Lucas, Nancy. "Robot Cop." *Omni* 6 (October 1984), 55.

Describes robot used by Arizona Department of Public Safety to cope with hostage situations and other dangerous confrontations, built by Bob Pederson of Toronto, Canada. Illustrated with photograph.

1288. Montgomery, Reid H., Jr., and MacDougell, Ellie C. "Curing Criminals." *The Futurist* 20 (January/February 1986), 36-37.

Includes discussion of uses of robots in prisons. Illustrated with drawing.

1289. Rifkin, Glenn. "Robot Security Guards: R2D2 on the Alert." *Computerworld* 19 (October 7, 1985), U9.

Describes security robot produced by Denning Mobile Robotics, Woburn, Massachusetts. Illustrated with photograph.

1290. Riggs, Larry. "Robots Go to Jail." *Sales & Marketing Management* 134 (March 1985), 53-54.
Reports on mobile robots being developed to use to patrol and monitor prisons, marketed by Southern Steel, San Antonio, Texas. Illustrated with photographs.

1291. Rogers, Michael. "Birth of the Killer Robots." *Newsweek* 103 (June 25, 1984), 51.
Describes developments in the robot security field. Reports on robot sentry, the Prowler, being developed by Robot Defense Systems of Thornton, Colorado, and robot prison guard developed by Denning Mobile Robotics in Woburn, Massachusetts. Also reports on military development of battlefield robots. Mentions that no robot should ever be given the ability to start its own army. Includes photographs.

1292. Wierzbicki, Barbara. "Mobile Robots Scheduled for Testing as Prison Guards." *InfoWorld* 6 (February 20, 1984), 17.
Describes patrol robots for medium security prisons that would carry no weapons but have closed-circuit television cameras and two-way radios.

Promotional Robots

1293. "Detroit Hits the Road with New Shopping Mall Shows." *Business Week* no. 2839 (April 23, 1984), 47.
Reports on the use of 24 robots that look human, used to promote Dodge cars at shopping center exhibits. Includes color photograph.

Sports Robots

1294. "RoboCaddy." *Popular Mechanics* 165 (December 1988), 13.
Reports on robotic golf caddie. Illustrated with color photograph.

1295. "Robot Carries Clubs for Walking Golfers." *Insight* 4 (May 2, 1988), 48.
Reports on robotic golf caddie, a wheeled vehicle which carries golf clubs and follows a transmitted signal from a transmitter carried by the golfer, developed by Gettig Technologies, Incorporated, Spring Mills, Pennsylvania. Includes photograph.

1296. "Robots in the Rink." *Popular Science* 236 (June 1990), 41.
Reports on Robo-Puck hockey tournament using student-built robots at Massachusetts Institute of Technology, Cambridge, February 7, 1990.

1297. "The Longest Ball." *Golf Magazine* 27 (February 1985), 42-43.
Describes test to determine the longest ball on the green, using Perfy, a putting robot designed by *Golf Magazine* technical consultant Dave Pelz. Illustrated with color photograph.

1298. "Wimbleton for Engineers." *Environment* 26 (May 1984), 24.
Reports on the organization of a Ping-Pong game between robots.

1299. Andersson, Russell L. *A Robot Ping-Pong Player: Experiment in Real-Time Intelligent Control*. Cambridge, MA: The MIT Press, 1988. 275 p.
Provides technical description and analyzes construction of the first robot to play Ping-Pong against humans and machines, developed at American Telephone & Telegraph Bell Laboratories. Illustrated with photographs.

1300. Bryant, Nelson. "Outdoors: Automatic Duck Adds Touch to Retrieving." *New York Times* 137 (September 21, 1987), C11.

Describes robot duck decoy that can retrieve downed ducks. Illustrated with drawing.

1301. Dworetzky, Tom. "I Got the Jock Right Here; He Sure Ain't Paul Revere." *Discover* 8 (June 1987), 14.

Reports on Super-Jock, a 22-pound robot jockey used on horses in indoor racing. Includes color photograph.

1302. Johnstone, Bob. "Direct-Drive Robots Play the Game." *New Scientist* 113 (March 5, 1987), 30.

Describes a robot built at Toshiba's Energy Science & Technology Laboratory, Kawasaki, Japan, that can play Ping-Pong. Illustrated with photograph-drawing combination.

1304. Lennard, Amy. "Robots on Board." *Sports Illustrated* 66 (January 26, 1987), 8.

Describes Super-Jock, a robot jockey used in live horse racing. Illustrated with color photograph.

1305. Neff, Craig. "And in This Corner, Herr Robot." *Sports Illustrated* 67 (July 6, 1987), 18.

Describes Der Boxroboter, a robot designed to be a sparring partner that can be programmed to fight in a variety of styles, developed in East Germany. Illustrated with color photograph.

1306. Skorupa, Joe. "Outdoors...Is It a Bird?" *Popular Mechanics* 165 (February 1988), 44.

Reports on Robo-Duck, a robot duck-retrieving decoy, developed by Outdoorsmen's Herters, Beaver Dam, Wisconsin. Illustrated with color photograph.

1307. Stefanides, E. J. "Self-Guiding Caddy Cart Carries Bags, Follows Golfer." *Design News* 44 (June 6, 1988), 136-138.

Describes the Shedda Electronic Golf Caddy, self-propelled, self-guided caddie, developed by Gettig

Engineering & Manufacturing Company, Spring Mills, Pennsylvania. Illustrated with color photograph and drawing.

1308. Zern, Ed. "Mechani-Mutt Revisited!" *Field and Stream* 91 (October 1986), 162.

A tongue-in-cheek description of a proposed robot hunting dog, update of a proposal first made in the April 1959 issue. Illustrated with drawing.

Underwater Robots

1309. "A 'Mr. Clean' for Offshore Platforms." *Design News* 43 (March 23, 1987), 32.

Describes a remotely operated cleaning and inspection system for use at offshore platforms, developed by Dawson Industries, Perth, Australia. Illustrated with color photographs.

1310. "Giving Robot a Helping Hand." *Design News* 43 (August 17, 1987), 58, 60.

Reports on research at Woods Hole Oceanographic Institute, Cape Cod, Massachusetts, to give an arm to remotely controlled underwater robot *Jason*. Illustrated with color photograph.

1311. "R2D2 of the Deep Debuts Swimmingly." *Changing Times* 37 (June 1983), 18.

Short description of an underwater robot, technically called a Remotely Piloted Vehicle System (RPV), but nicknamed R2D2 of the Deep. The robot was designed by Chris Nicholson, who built his first underwater vehicles for a high school science fair. It is used to check the condition of underwater oil rigs and pipelines, and for deep-water recovery missions. Includes photographs.

1312. "Remote Control Checks Pipelines for Safety." *New Scientist* 118 (May 19, 1988), 40.

Reports on remotely controlled device used with remotely operated vehicle to check underwater pipelines for safety. Illustrated with photograph.

1313. "Robot for Underwater Construction." *The Futurist* 22 (November/December 1988), 39-40.

Reports on Aquarobo, an underwater robot that walks on six legs, being developed by the Port and Harbour Research Institute of Yokosuka City, Japan. Illustrated with photographs.

1314. "Underwater Robots." *Economist* 290 (October 29, 1983), 92.

Reports on uses and development of underwater robots.

1315. Ballard, Robert D. "A Long Last Look at *Titanic*." *National Geographic* 170 (December 1986), 698-727.

Includes description of how underwater robot *Jason Jr.* was used to explore the *Titanic*. Illustrated with color and black-and-white drawings and photographs.

1316. Bower, Bruce. "New Look at *Titanic* with a Robot's 'Eye.'" *Science News* 130 (July 19, 1986), 37.

Describes use of the camera-bearing robot *Jason Jr.* to explore and photograph the *Titanic*. Includes photograph.

1317. Britton, Peter. "Robot Subs Trouble-Shoot the Deep." *Popular Science* 219 (December 1981), 70-72.

Reports on uses of underwater robot subs. Includes color drawings. Cover of this issue illustrates a working underwater robot sub.

1318. Britton, Peter. "Submersible Acrobat." *Popular Science* 234 (April 1989), 96.

Describes Scorpio Scout, a remotely operated vehicle (ROV) for underwater use, invented by Ted Haselton and produced by Ametek of San Diego, California. Illustrated with color drawing and photograph.

1319. Britton, Peter. "The New Generation of Brainy Robot Subs." *Popular Science* 223 (December 1983), 102-105.

Discusses the use of remotely operated vehicles (ROVs) called autonomous underwater vehicles (AUVs), which are usually untethered or have only a modified tether to a mother ship. Describes an epic dive by *Epaulard*, French AUV that located and photographed a World War II bomber on the floor of the Pacific Ocean near San Diego, California. Illustrated with color drawings and painting.

1320. Budiansky, Stephen. "Beyond *Titanic*: Charting the Deep." *U. S. News & World Report* 101 (August 18, 1986), 56-57.

Reports on uses of robots with underwater cameras, especially for exploration of the ocean bottom. Illustrated with color photographs and drawings.

1321. Connor, Steve. "How to Rid the Gulf of Mines." *New Scientist* 115 (August 20, 1987), 26.

Discusses Great Britain's use of glass-reinforced, plastic ships in Persian Gulf, equipped with robot submersibles that locate mines underwater, photograph them, and detonate them.

1322. Emanuel, Dave. "Robot Repair Crew for 7,000-Ft.-Deep Pipeline." *Popular Science* 222 (January 1983), 55-57.

Describes plan to use robots (remotely controlled vehicles) to repair a 100-mile-long natural-gas pipeline running from Spain to Algeria. Includes color drawings and painting.

1323. Francis, Ted. "Underwater Robot." *Popular Science* 232 (April 1988), 122.

Discusses the use of robot submarines. Illustrated with photograph.

1324. Holden, Constance. "Americans and French Find the *Titanic*." *Science* 229 (September 27, 1985), 1368-1369.

Reports on the finding of the *Titanic* by an expedition of scientists from the Deep Submergence Laboratory at Woods

Hole Oceanographic Institute, Cape Cod, Massachusetts, using pictures sent by underwater robot *Argo* after a French expedition found the immediate area.

1325. Lamb, John. "Eureka Projects Keep on Coming." *New Scientist* 115 (September 24, 1987), 33.

Describes robot designed to clean and maintain oil rigs and other structures in the sea and robot that will be able to travel away from its base, locate an area of the sea bed, and survey both the surface and deposits buried beneath it.

1326. Lemonick, Michael D. "'J. J.' Tours the *Titanic*." *Time* 128 (July 28, 1986), 52.

Reports on the underwater robot *Jason Jr.*'s exploration of the *Titanic*. Includes color photographs.

1327. Marbach, William D., with Conant, Jennet, and Lord, Mary. "ROVing the Ocean Floor." *Newsweek* 102 (September 26, 1983), 86.

Reports on developments in remotely operated underwater vehicles (ROV's). Describes underwater robot EAVE (Experimental Autonomous Vehicle), equipped with onboard computers, sonar, and primitive artificial intelligence, programmed to navigate an underwater course and make its own decisions about turning, avoiding objects, or returning to its starting location. EAVE was built by the Marine Systems Engineering Lab at the University of New Hampshire, Durham. Also reports on ROV's being developed at Woods Hole Oceanographic Institute in Woods Hole, Massachusetts.

1328. Meier, Barry. "Robot Subs Begin to Surface as Versatile Exploration Tools." *The Wall Street Journal* 207 (March 7, 1986), 25.

Describes uses of underwater robot subs.

1329. Meigs, James B. "Exploring with Video." *Popular Mechanics* 163 (November 1986), 22, 27.

Describes video cameras used by Dr. Robert Ballard and team of scientists who explored the *Titanic*, using robots *Alvin* and *Jason Jr.* Illustrated with photographs.

1330. Murphy, Jamie, et al. "Down into the Deep." *Time* 128 (August 11, 1986), 48-54.
Reports on underwater explorations, including use of underwater robot *Jason Jr.* in exploration of the *Titanic*. Illustrated with color photographs and drawings.

1331. Powell, Stewart. "Journey Through the Portholes of a Gilded Past." *U. S. News & World Report* 101 (July 28, 1986), 8.
Reports on the *Titanic* video pictures taken by the remote videocamera robot *Jason Jr.* Includes photograph of the *Titanic*'s chandelier.

1332. Scanlon, Kevin. "Exploring the *Titanic*'s Watery Grave." *Maclean's* 99 (July 28, 1986), 46.
Reports on findings and use of underwater robot *Jason Jr.* as well as findings of expedition divers exploring the *Titanic*. Includes photographs.

1333. Scanlon, Kevin, and Lowther, William. "Revising the *Titanic*'s Popular Legend." *Maclean's* 99 (August 11, 1986), 42.
Reports on the use of underwater robot *Jason Jr.* to videotape the *Titanic* and future uses of underwater robots. Includes photograph.

1334. Shaker, Steven M. "Robot Spy Sub." *Popular Science* 232 (February 1988), 62.
Reports on Night Surveillance System, videocamera-equipped, semi-submersible robot built by Perry Offshore, Incorporated, Riviera Beach, Florida, for use in waterside security and illicit shipping detection. Illustrated with color photographs.

1335. Swann, Christopher. "Robot Mini-Subs Replace Pro Divers." *Popular Science* 227 (November 1985), 80-81.

Describes several remotely operated vehicles (ROVs) for underwater use, and how they are used. Illustrated with color photographs.

1336. Tucker, Jonathan B. "Submersibles Reach New Depths." *High Technology* 6 (February 1986), 16-21, 23-24.

Provides in-depth coverage of developments involving remotely operated vehicles (ROVs), underwater robots. Includes photographs and a drawing.

Other Applications

1337. "An 'Intelligent' Robot." *USA Today* 109 (December 1980), 11-12.

Reports on a robot with wheels, equipped with television camera and transmitter, with a number of possible applications such as Mars rover or submersible rover. The robot was developed by Hans Moravec at Stanford University's Mechanical Engineering Department and Artificial Intelligence Laboratory, Stanford, California. Illustrated with photograph of the robot and Moravec.

1338. "Current Market Potential for Mobile Robots Near 100,000, Study Reports." *Industrial Education* 72 (October 1983), 17, 46.

Discusses possible applications for mobile robots.

1339. "Dummies Get Smarter." *Popular Mechanics* 165 (December 1988), 33.

Reports on ADAM, robot ("intelligent mannequin") used to test pilot ejection seats; Manny, robot used to test protective clothing; and Eddie Endo, robot used for practicing administration of anesthesia. Illustrated with color photographs.

1340. "Enter the Robot Librarian." *New Scientist* 97 (March 3, 1983), 586.

Reports on "intelibots," battery-driven, wheeled robots the size of shoeboxes, used to pick out videotapes from stocks and put them into players at Kanazawa Industrial University Library in Japan.

1341. "Free-Wheeling Plotter Roams Over Drawing." *Machine Design* 57 (April 11, 1985), 60-61.

Describes a robot plotter that produces multi-color drawings of personal computer graphics, developed by Penman Products, Limited, West Sussex, England. Illustrated with photographs and drawings.

1342. "Heath Robinson Robots." *The Economist* 279 (June 27, 1981), 97.

Reports on robot applications such as guard dog, robot patient, robot nurse, and robot waitress. Illustrated with drawing.

1343. "It's Alive! (Or Is It?)." *National Geographic World* 143 (March 1989), 29-33.

Describes robotic dinosaurs made for museums by Dinamation International Corporation, California. Illustrated with color photographs.

1344. "Jackknife Carloader." Design News 43 (July 20, 1987), 51-53.

Describes robot used to lift automobiles into an enclosed truck and place them in designated positions, designed by Dallis Smith Engineering Company and built by Transportation Manufacturing Corporation, a division of Greyhound Corporation. Illustrated with color and black-and-white photographs and color drawings.

1345. "Parapeople." *Horizon* 20 (November 1977), 75.

Reports on "androids," robots developed by Quasar and used for business promotions, homes, and experimentally with hospital patients. Photograph shows a robot named Klatu walking a dog.

1346. "Robo-Mower." *The Futurist* 23 (January/February 1989), 39.
Describes a robotic lawn mower developed by Jim Hammond and Ray Rafaelo of Technical Solutions, Incorporated, Sterling, Virginia. Illustrated with color photograph.

1347. "Robot 'Sweats Out' Hazardous Environments." *Design News* 44 (August 1, 1988), 34.
Reports on Manny, robotic mannequin being developed to test protective clothing in dangerous environments by Battelle's Pacific Northwest Laboratories, Richland, Washington. Illustrated with color photograph.

1348. "Robot to the Rescue." *Science Digest* 72 (December 1972), 74.
Reports on a firefighting robot in Yokohama, Japan. Illustrated with photograph.

1349. "Robotic Lineman." *Popular Mechanics* 165 (March 1988), 22.
Reports on TOMCAT (Teleoperator for Operations, Maintenance and Construction Using Advanced Technology), a robot for use in overhead transmission line repair designed by Kraft Oceanic Systems of Philadelphia, Pennsylvania. Illustrated with color drawing.

1350. "Robotic Owl Performs Realistic Flight Routines." *Machine Design* 57 (October 10, 1985), 42.
Describes a robot snowy owl with inflight maneuvers and perching movements, used in an exhibit designed for museums. Illustrated with drawing.

1351. "Robotics Institute Readies REX, Rover & Co." *Carnegie-Mellon Magazine* 2 (Winter 1984), 18-19.
Reports on robots being developed at Carnegie Mellon University, Pittsburgh, Pennsylvania: the Ferret for inspecting pipes and conduits; the Robot Excavation System, to look for leaking utility lines; the Turtle, to enter damaged

nuclear reactors; and the Rover and other robots designed to walk or move over the environment.

1352. "Tech Update." *Popular Mechanics* 164 (August 1987), 96.

"Robot Guard Duty" describes sentry robot manufactured by Denning Robotics, Woburn, Massachusetts, and includes drawing. "Automated Plane Builder" describes a robotic carousel head used to fasten plane wings together, manufactured by Taylor-Hitec, Limited, Great Britain, and includes photograph. "Robot Biped" describes a biped robot built by Yuan Zheng of Clemson University, Clemson, South Carolina, and includes photograph.

1353. "The Talk Circuit; Presenting Robot Redford." *Time* 121 (May 9, 1983), 39.

Describes plans for a robot to deliver the commencement address at Anne Arundel Community College in Maryland.

1354. "Wall-Crawling Robot." *Popular Mechanics* 165 (June 1988), 17.

Reports on hand-sized robot called The Crab which can climb walls and negotiate curves, designed by French defense contractor CRAM.

1355. "When You're a Mom." *'Teen* 30 (December 1986), 26, 28.

"Robot Nannies" section describes what household robots may do in the future. Illustrated with drawing.

1356. "Your Wish Is Robot's Command." *Design News* 43 (March 23, 1987), 44.

Reports on the six-axis RTX, claimed to be the "first personal working robot," developed by Universal Machine Intelligence Company, Camberly, Surry, England, for use in areas such as laboratory experiments and to aid the disabled.

1357. Ahl, David H. "Dr. Kato and His Amazing Robot Friends." *Creative Computing* 10 (August 1984), 115-116.

Reports on the work of Professor Ichiro Kato with a piano and organ playing robot (WAM-7R), a biped walking robot, and a robotic leg prosthesis. Includes color photographs.

1358. Berkstresser, Gordon A., III, and Buchanan, David R., eds. *Automation and Robotics in the Textile and Apparel Industries*. Park Ridge, NJ: Noyes Publications, 1986. 328 p.

Includes robotics overview, machine vision, tactile sensing, robotic materials handling, automated sewing systems, sewing industry research, potential uses for robots in textile and related industries, need for automation, directions of technological change in fiber, textile, and apparel industries, corporate planning issues, textile and apparel technology in the United States, and American society in the age of advanced automation. Illustrated with photographs and drawings.

1359. Brown, Stuart F. "Meet the Wallbots." *Popular Science* 233 (August 1988), 54-55.

Describes Trackbot and Studbot, robot-like machines designed to aid in large building construction, being developed at Massachusetts Institute of Technology's Center for Systems Automation in Cambridge. Illustrated with color photographs and drawings.

1360. Bukowski, Norbert. "Starting a Robotics Company." *Micro Cornucopia* no. 47 (May-June 1989), 64-66.

Relates the history of Bukowski Robotics, Incorporated, Tempe, Arizona, makers of robots for various commercial uses. Illustrated with photographs.

1361. Bylinsky, Gene. "Get Ready for a Robot Butler." *Reader's Digest* 133 (January 1988), 157-158, 161-162.

Reports on uses and possible applications of service robots in medicine, nuclear plants, underwater activities, the military, and other areas, and on work being done by Joseph F. Engelberger's Transitions Research Corporation, Danbury,

Connecticut. A condensed version of *Fortune* article, number 1362.

1362. Bylinsky, Gene. "Invasion of the Service Robots." *Fortune* 116 (September 14, 1987), 81-82, 84, 86, 88.
Describes ways robots can be used or are being used in businesses and services outside of industry. Includes color photographs.

1363. Chapman, Bert. "Robots Leave the World of Sci-Fi to Enter Graphics Communications." *Graphics Arts Monthly* 53 (August 1981), 37-42.
Reports on ways industrial robots are used, including applications in the newspaper and printing industries. Illustrated with color and black-and-white photographs. Cover of this issue shows an industrial robot in color.

1364. Coatar, Joseph. "The Mail Machine." *Management World* 11 (June 1982), 36-37.
Describes the Mailmobile, a robot vehicle to deliver mail in a large, 18-floor office. Illustrated with photograph.

1365. Cornish, Edward. "'Did You Hear the One About the Human Who ...'" *Computerworld* 20 (November 3, 1986), S28-S29.
Discusses uses of home robots in the future such as robotic furniture (voice-commandable tables and chairs), voice-commandable robotic closets, floor-cleaning robots that can ask a sofa to move, with the sofa responding by moving, and television used with a robot to transmit signals over distance to a human hand. Illustrated with drawing.

1366. Davis, Brian. "Robotics on the High Seas." *The Engineer* 256 (May 19, 1983), 30.
Describes robot use in shipbuilding in England.

1367. Day, Charles R., Jr. "Robots Man the Mail Carts; but at Zale Corp. They Don't Look Like Robots." *Industry Week* 223 (November 12, 1984), 113-114.

Describes the Transcar system, a West German-built, robot delivery system installed by Lamson Corporation, Syracuse, New York, at Zale Corporation, Dallas, Texas. Illustrated with photographs.

1368. Eisenstein, Paul A. "The Future of the Robot." *Industrial Education* 77 (October 1988), 26-27.

Discusses ways service robots are used now and might be in the future. Illustrated with photograph.

1369. Fisher, Arthur. "Sweaty Manny." *Popular Science* 233 (September 1988), 10, 12.

Reports on Manny, robotic mannequin being developed to test protective clothing in dangerous environments by Battelle's Pacific Northwest Laboratories, Richland, Washington. Illustrated with color photograph.

1370. Fisher, Thomas. "[P/A Technics: Robotics in Japan.]" *Progressive Architecture* 67 (June 1986), 110-111.

Discusses use of robots in the construction industry, especially in Japan. Illustrated with drawings and color photographs.

1371. Frand, Erwin. "Some Terrific New Product Ideas." *Research & Development* 29 (July 1987), 19.

Discusses uses for an artificially intelligent robot, yet to be developed, that could be programmed to act like a baby and a child to give child-rearing experiences to future parents.

1372. Hanna, Annetta. "It's a Bird, It's a Plane..." *Industrial Design* 33 (March/April 1986), 50-51.

Describes design and flight of robot pterodactyl, a giant prehistoric flying reptile. Illustrated with photographs.

1373. Harrar, George. "The Service Robot Lumbers Off the Drawing Board." *New York Times* 139 (August 12, 1990), 139.

Reports on ways service robots are being used or being considered for use, especially in restaurants, hotels, hospitals, nuclear plants, sentry duty, and the postal service. Illustrated with photograph.

1374. Hawkins, William J. "For the Active Couch Potato." *Popular Science* 232 (June 1988), 19.

Reports on Animan, a walking television set designed by Brian Elliott while a student at the Art Center College of Design, Pasadena, California. Illustrated with color photograph.

1375. Horgan, John. "Disco-Bee." *Scientific American* 260 (June 1989), 31-32.

Reports on computer-controlled robot honeybee that performs bee-dance steps, developed by Wolfgang H. Kirchner of the University of Würzburg, West Germany, and Axel Michelsen of Odense University, Denmark.

1376. Jorden, William H., Jr. "They Move, They Roar: Dinosaurs Are Here Once More." *Smithsonian* 20 (August 1989), 46-52, 54, 56-57.

Reports on Dinamation International Corporation in Orange County, California, makers of robot dinosaurs and other prehistoric animals used in museums. Illustrated with color photographs.

1377. Kohler, Pat. "Robots, Like Software, Are Best Taken to Task in Office Environment." *PC Week* 2 (January 29, 1985), 45, 47.

Discusses ways robots could be used in an office.

1378. Lee, Shirley. "Monster Mechanic." *Life* 72 (November 1982), 26-27, 72.

Describes work at a special effects and animated figure company, AVG Productions, Valencia, California. Illustrated with photographs.

1379. Liversidge, Anthony. "Space Pets." *Omni* 7 (October 1985), 42.

Discusses human need for pets in space and possible use of robots. Illustrated with photograph.

1380. Marbach, William D. "This Steel-Collar Seamstress Has a Velvet Touch." *Business Week* no. 3098 (April 3, 1989), 110.

Reports on two-armed robot to be used on automated sewing lines in a clothing factory, developed by Matsushita Electric Industrial Company in Japan. Illustrated with color drawing.

1381. Matthews, Robert. "Slave to the Market." *Design* no. 457 (January 1987), 30-31.

Describes robot arm, RTX, for disabled, universities, and small companies, built by Tim Jones, co-founder of Universal Machine Intelligence, Wandsworth, England. Illustrated with photographs.

1382. Moffett, Mark W. "Dance of the Electronic Bee." *National Geographic* 177 (January 1990), 134-140.

Describes work on robot bees being conducted by scientists at Odense University in Denmark, and the University of Würzburg in West Germany. Illustrated with color photographs and drawings.

1383. Myers, Frederick S. "Play It Again, Wabot." *Scientific American* 262 (May 1990), 84-86.

Reports on robots, including piano playing robot, being developed at Waseda University, Tokyo, and the Tokyo Institute of Technology in Japan. Illustrated with color photograph.

1384. Norton, James. "Robots Unlock Secrets of Child Sexual Abuse." *The Police Chief* (April 1987), 31-32, 36.

Describes modified ARES robot used by Valdez Department of Emergency Services, Valdez, Alaska, to interview children who are victims of sexual abuse. Illustrated with photographs.

1385. Schmidt, Klaus F. "Robotics in the Graphic Arts." *Print* 38 (January/February 1984), 103-104.
Reports on industrial robots and potential in the graphic arts industry.

1386. Scott, David. "Robot Driver Racks up the Miles for Lab Tests." *Popular Science* 228 (April 1986), 40.
Describes robot driver built by Motor Industry Research Association of Great Britain, to test-drive vehicles.

1387. Scott, David. "Robot Plotter for Low-Cost Graphics." *Popular Science* 225 (November 1984), 120.
Describes three-color, turtle-like robot plotter for use with a computer to draw charts, schematics, and other hard copy graphics, available from Chessell Microproducts, Worthing, West Sussex, England.

1388. Serdahely, Franz. "Now Computers and Robots Join the Fight Against Fire." *Popular Science* 223 (November 1983), 50, 52, 54, 58.
Includes photograph and information on Snail, a robot that carries water lines into fires, created by the Scottsdale, Arizona, Rural-Metro Fire Department.

1389. Siwolop, Sana. "Of Micromice and Men: the U. S. Tries to Build a Better 'Rodent.'" *Business Week* no. 2972 (November 10, 1986), 130.
Reports on micromouse competition in which little robots run a maze, with the winner being the robot that finds the fastest path to the middle of the maze. Includes color drawing.

1390. Vaughan, Don. "Robot Recharges Museum." *High Technology* 3 (December 1983), 10.
Describes Sir Plus, a robot used at the Palm Beach (Florida) Science Museum as a guide. Includes information on ways the robot was designed to make it safe with children. Illustrated.

1391. Warszawski, Abraham. *Industrialization and Robotics in Building*. New York: Harper and Row, 1990. 466 p.
On the use of prefabrication and robots for building construction. Includes sections on robot technology, building robotics configuration, performance, and implementation aspects. Illustrated with photographs and drawings.

1392. Weiss, Rick. "New Dancer in the Hive." *Science News* 136 (October 28, 1989), 282-283.
Reports on computer-controlled robot honeybee that performs bee-dance steps, developed by Martin Lindauer and Wolfgang H. Kirchner of the University of Würzburg, West Germany, and Axel Michelsen of Odense University, Denmark. Illustrated with drawing.

1393. Wilken, Earl. "Robots Are Headed for Newspaper Mailrooms and Reelrooms." *Editor and Publisher* 114 (February 14, 1981), 35, 41.
Reports on uses of robots in newspaper and publishing industries. Includes photographs and drawings.

1394. Wong, Jan; Yoder, Stephen Kreider; and Hudson, Richard L. "Robots: Next Step for Garment Makers." *Wall Street Journal* 208 (August 7, 1986), 22.
Describes robots and automated equipment being developed or planned for use in the garment industry.

1395. Wouk, Henry. "Shopping by Robot." *Omni* 8 (June 1985), 140.
Describes ways robots are used in a supermarket in Japan. Illustrated with photograph.

1396. Yates, David. "Let Your Robot Do the Walking." *Science Digest* 91 (May 1983), 30, 32, 34.
Describes a proposed robotlike device, called a "proxy," under a person's control, that can see, hear, and move, and that would pass on to the controller the images and sounds it encounters. The robot is intended to give the controller the sensory impression of being present in the remote

location. Discusses how it might be built, some applications, and possible problems and consequences. Includes drawings.

SECTION 6. SOCIAL IMPLICATIONS OF ROBOT USE

The social implications of the use of robots have been a favorite topic of speculation in both nonfictional and fictional writings. No one knows for sure what a real robot revolution might lead to in our civilization. Some fear that our ultimate fate will be as slaves to master robots rather than the other way around.

Nonfictional materials often deal with studies conducted to determine the impact on humans of robot introduction into a particular workplace or environment. The robot is frequently seen as a problem as well as a solution. Feelings are mixed. Distrust of the robot, even though it is essentially a piece of machinery rather than a quasi-human entity, leads to sabotage of robots and their productivity. Perceived values often fail to include the robot's usefulness to the quadriplegic, or for working in extremely dangerous environments (e.g., nuclear reactors) and in space exploration. There is sometimes a foreboding that the robot will get out of hand and, by self replication, destroy humanity.

Some materials in this section deal with the idea of civil rights for robots, giving them the same rights to existence and fair treatment as humans have. The idea of working with robots as equal partners is also considered. Students will find these materials provocative and thought-producing.

1397. Albus, James S. "The Federal Role in Robotics." *Enterprise* 6 (August 1982), 5-9.

Discusses industrial robots, robot-automated factory, worker and economic implications, government standards, policymaking and other roles, prospects, and problems. Illustrated with photograph.

1398. Asimov, Isaac. "The Robots Are Coming." *Consumers Digest* 24 (March/April 1985), 84-86.

Discusses types of work robots will do, kinds of worker jobs that will be required involving robots, possible changes in work and the society, problems of intelligent robots, and future computer/robot/human being creativity. Illustrated with drawing.

1399. Ayres, Robert U., and Miller, Steven M. *Robotics: Applications and Social Implications*. Cambridge, MA: Ballinger Publishing Company, 1983. 339 p.

Examines the social implications of the use of robots, particularly in industry. Chapters deal with robot/CAM technology, costs and benefits of robots in manufacturing, applications of robots in hazardous or inaccessible environments, employment impact of robotization, productivity impacts, policy implications, and the ultimate robot. An earlier version is listed in ERIC document ED 231 947 as *The Impacts of Robotics on the Workforce and Workplace*.

1400. Chamberlin, Leslie J. "Facing up to Robotation." *USA Today* 111 (November 1982), 31-33.

Discusses the impact of industrial robots, including why industrial robots are used and how they will change work in industry, school programs, and our way of life. Illustrated with drawing.

1401. Coates, Vary T. "The Potential Impacts of Robotics." *The Futurist* 17 (February 1983), 28-32.

Considers changes robots can make in industry, employment, and, eventually, society, based on the Technology Diffusion or User-Institution Model, which shows how an institution responds to a new technology, going through the stages of substitution, accommodation, and innovation. Illustrated with photographs.

1402. Frude, Neil. *The Intimate Machine; Close Encounters with Computers and Robots*. NY: New American Library, 1983. 244 p.

Considers human reaction to and interaction with computers and robots now and in the future.

1403. Joseph, Earl C. "Ask a Futurist. Peace (and) Robots." Paper presented at the World Future Society's 5th Assembly, Washington, DC, July 1984. ERIC document 245 971.

Discusses the future of robots and the difference between robots and expert systems in creating new jobs and raising society's standard of living.

1404. Lesser, Simon O. "Our Feeling About Man-Made Creatures, Imaginary and Real." *Literature and Psychology* 15 (Winter 1965), 28-31.

Comments on "The Golem and the Robot" by Robert Plank, and on rational and irrational human feelings about human-made creatures.

1405. Ludden, Laverne; Roudebush, Deborah; and Weaver, Roy A. "The Interdependence of Computers, Robots, and People." Paper presented at the National Adult Education Conference, Philadelphia, Pennsylvania, December 1, 1983. 44 p. ERIC document ED 237 659.

Discusses problem of providing jobs for workers displaced by computers and robots, impact on adult education programs, what should be done with leisure time created by this technology, and at what point robots' intelligence and possible superiority over people should be acknowledged.

1406. McConkey, Dale, and Barrett, F. D. "Managing in the Age of the Robot." *Business Quarterly* 47 (December 1982), 40-46.

Discusses replacement of employees with robots, how robots will change management ratios, practices, and skill requirements, and social issues to consider.

1407. McNally, Phil, and Inayatullah, Sohail. "The Rights of Robots." *Whole Earth Review* (Summer 1988), 2-10.

Discusses rights of nature, such as trees, robot technology, possibility of "life" for robots, defining rights, emergence of rights, economic issues, new court cases involving robots, the electronic judiciary, and legal principles. Illustrated with photographs and drawings.

1408. Miller, Robert J. "The Human: Alien in the Robotic Environment?" *Annals. American Academy of Political and Social Science* 470 (November 1983), 11-15.

Considers the robot and human environments, evolving robot behavior and human behavior, ramifications in the workplace and beyond, and the man-machine relationship.

1409. Moore, Dan T., Jr. "Will Robots Save Democracy?" *The Futurist* 15 (August 1981), 14-19.

Discusses warning signs of the decline of democracy and how robots may affect society in the future. Illustrated with photographs and drawings.

1410. Moore, Oliver S., III. "The Robots Are Coming." *Science Digest* 93 (July 1985), 4.

Presents an editorial on robots and society.

1411. Parsons, H. N., and Kearsley, G. P. "Robotics and Human Factors: Current Status and Future Prospects." *Human Factors* 24 (October 1982), 535-552.

Examines human engineering factors. Considers manufacturing, institutions, space, underseas, mining, agriculture, construction, home, nuclear plants, and other applications.

1412. Simons, Geoff. *Are Computers Alive?* Cambridge, MA: Birkhauser Boston, Inc., 1983. 212 p.

Covers question of the computer and robot with artificial intelligence as an emerging life form, behavior of computers and robots, robot anatomy, psychology of computers, computer liberation, the human response, civil rights for computers and robots, and the future.

1413. United States Congress. Office of Technology Assessment. *Exploratory Workshop on the Social Impacts of Robotics. Summary and Issues. A Background Paper.* 1982. 142 p.

Discusses future of industrial robots and likely impact on public policy, including social issues of productivity and capital formation, labor, education and training, and international imports, and includes papers on industrial robots.

1414. Warnat, Winifred I. "Human Factors Beyond the Workplace: A Social Context for Robotics." Paper presented at the Second Canadian Computer Aided Design/Computer Aided Manufacturing and Robotics Conference, Toronto, Ontario, Canada, June 1, 1983. 19 p. ERIC document ED 268 974.

Considers current and future employment picture, robotics industry developments and their impact on the changing character of work, and implications of changes for society and the worker.

1415. Zeldman, Maurice. "The Myth and Reality of Robots." *SA>M Advanced Management Journal* 49 (Autumn 84), 30-36.

Discusses the society implications of industrial robot use, and what management, labor, and government can do together to minimize any problems. Illustrated with photographs.

SECTION 7. ECONOMIC IMPLICATIONS OF ROBOT USE

This section deals with the economic ramifications of the robot, particularly the robot's impact on employment. Workers whose replacement jobs are at a lower pay scale than jobs held before robots took over their jobs are likely to fall economically behind. In company towns, regions, or countries where job replacement occurs in a large number of companies or throughout an entire industry (such as the automobile industry), economic hardships are certain to occur. Highly skilled workers who previously received substantial salaries may find their skills outdated and unneeded. On the other hand, the same companies or industries, by saving money with the more economical robots, may be able to amass considerable sums of money. One author considers the problems of wealth distribution in a future robot economy. Studies in Michigan and Japan consider the impact of industrial robots on jobs.

Materials listed in this section will provide background for class discussions and reports on what robots are doing to economies today and how they might affect employment worldwide in the future.

1416. "Robotics Study Projects Modest Impact on Employment." *Editor & Publisher* 116 (January 22, 1983), 36.

Reports on results of study, "Robotics: Human Resource Implications for Michigan," which considered the impact of robots on jobs, conducted by W. E. Upjohn Institute for Employment Research, Kalamazoo, Michigan.

1417. "The Robot Invasion Begins to Worry Labor." *Business Week* no. 2732 (March 29, 1982), 46-47.

Reports on a planned Japanese government study on the effect of robots on employment, and reactions of labor unions in Japan to industrial robot use. Includes photograph.

1418. Albus, James S. "Robots and the Economy." *The Futurist* 18 (December 1984), 38-44.

Discusses the implications of the next industrial revolution, in which computer power will be substituted for brain power in the control of machines and industrial processes, and which will be based on robot labor. Suggests that the single most important barrier to the new industrial revolution is the unanswered question of unemployment and competition for jobs between robots and human workers. Considers reasons for the fear of robots and gives some possible solutions to the problem of wealth distribution in a future robot economy. Includes photographs and other illustrations.

1419. Hunt, H. Allan, and Hunt, Timothy L. *Human Resource Implications of Robotics*. Kalamazoo, MI: The W. E. Upjohn Institute for Employment Research, 1983. 207 p.

Discusses what industrial robots are and how they are used, forecasts of the robot population, job displacement effects of robots, and job creations due to robots. Studies robot use in Michigan in particular. Includes annotated bibliography.

1420. Jenkin, Patrick. "'The Unemployed Cannot Blame Automation.'" *New Scientist* 97 (February 24, 1983), 526-527.

Addresses concerns that robots and automation caused unemployment in England, and gives reasons why this was untrue. Includes photographs.

1421. Levitan, Sar A., and Johnson, Clifford M. "The Future of Work: Does It Belong to Us or to the Robots?" *Monthly Labor Review* 105 (September 1982), 10-14.

Discusses possible impact of robots and other automated equipment on the workforce, pace at which technological change may occur, and reasons for a slower rate of change than predicted.

1422. Skrzycki, Cindy. "Will Robots Bring More Jobs - or Less?" *U. S. News & World Report* 95 (September 5, 1983), 25.
Discusses impact of industrial robots on the job market. Illustrated with color photograph and graph.

1423. Tanner, William R. "Robots Replacing Humans? A Displaced Concept." *American Metal Market* 92 (September 10, 1984), P.
Discusses displacement of workers by industrial robots.

1424. Turner, Steve. "When the Robots Punch In." *The Progressive* 47 (August 1983), 22-24.
Discusses possible effects of robot installations on jobs and job availability.

1425. United States Congress. House. Committee on Small Business. *Impact of Robots and Computers on the Workforce of the 1980's. Hearings Before the Subcommittee on General Oversight and the Economy of the Committee on Small Business*, 98th Cong., 1st sess., 1983. 259 p. ERIC document ED 244 120.
Presents testimony and statements on the impact of automation, including robots, on employment, and on retraining programs for displaced workers.

1426. United States Congress. Joint Economic Committee. *The Impact of Robotics on Employment. Hearing Before the Subcommittee on Economic Goals and Intergovernmental Policy of the Joint Economic Committee*, 98th Cong., 1st sess., 1983. 35 p. ERIC document ED 236 408.
Testimony by John Andelin stresses that robots are one component of a larger programmable automation phenomenon in industry, specific employment impacts are hard to predict, changes in number of jobs are only one of the consequences of increasing use of programmable automation, and curriculum development, change, and

delivery are not meeting changing industry needs. Testimony by Robert U. Ayres discusses the potential for, and problems of, future job displacement by robots.

SECTION 8. WORKPLACE IMPACT OF ROBOT USE

Materials in Section 8 focus on the impact on the worker when robots are brought into the workplace. Workers have reason to be concerned about the advent of the "steel collar worker" in their midst. One author discusses the robot as "colleague" and possibly "supervisor" of the worker. With little past experience to draw upon, workers have not yet adjusted themselves to this new possibility. Labor unions have stepped in to demand worker rights when faced with robot helpers or displacement by robots.

Worker displacement can be countered with job retraining. Robots can be brought into the workplace with less worker stress if workers are kept informed of progress in robotics implementation and how it will affect them. Several of the items in this section discuss how to introduce robots into the workplace with an emphasis on reducing negative worker impact.

Students will find this topic timely as they consider their own impending entry into the workplace and their feelings about working alongside robots. Ideas in the materials listed should challenge them to look for creative answers to the problems of working with, and perhaps for, robots.

1427. "It Is Harder to Find a Job Than to Do It." *Economist* 290 (March 24, 1984), 71-72.

Reports on the use of robots and computers by a number of countries, reactions of workers, and how to counter worker fear and distrust. Illustrated with photograph and graphs.

1428. "Robots: The Price of Progress." *Training* 20 (June 1983), 86.
Discusses worker impact, including job loss and retraining, when robots are introduced into factories.

1429. "Robots and Job Loss." *The Futurist* 19 (April 1985), 73.
Excerpts "Jobs for the Nation: Challenges for a Society Based on Work," by Anthony Patrick Carnevale for the American Society for Training and Development, which states that worries over worker displacement by robots in the future are unfounded. Illustrated with photograph.

1430. Amram, F. M. "Designing a Social Environment for Human-Robot Cooperation." In *Proceedings of the 1st International Conference on Human Factors in Manufacturing, 3-5 April 1984, London, UK*, pp. 127-136. Bedford, England: IFS (Publications) Ltd., 1984. Also available as ERIC document ED 241 969.
Discusses the robot as "colleague" and, potentially, as "supervisor" of the worker.

1431. Argote, Linda; Goodman, Paul S.; and Schkade, David. "The Human Side of Robotics: How Workers React to a Robot." *Sloan Management Review* 25 (Spring 1983), 31-41.
Reports on a study to examine workers' reactions to the introduction of a robot in one factory and changes in jobs brought about by the introduction, and presents strategies for introducing robots in the factory.

1432. Bruno, Charles. "Labor Relations in the Age of Robotics." *Datamation* 30 (March 1984), 179-180, 182.
Discusses introduction of manufacturing technology and tension between labor unions and management.

1433. Byham, William C. "HRD and the Steel-Collar Worker." *Training* 21 (January 1984), 59-60, 62, 64.
Discusses industrial robot implementation stages, situations where robotic installations did and did not work well, human problems, employment guarantees, roles for

human resource development professionals, and implications for workers. Illustrated with photograph.

1434. Chao, Georgia T., and Kozlowski, Steve W., Jr. "Employee Perceptions on the Implementation of Robotic Manufacturing Technology." *Journal of Applied Psychology* 71 (February 1986), 70-76.

Discusses a study of employee concerns about general robotics orientation, job security, management concern, and expected change.

1435. Cole, P. A. "Vocational Education and the Robotics Revolution." *GAO Review* 18 (Spring 1983), 21-25, 53-54.

Considers the possible impact on workers of robots in industry, and training programs for displaced workers and for future requirements. Illustrated with photographs.

1436. Crocker, Olga L., and Guelker, Richard. "The Effects of Robotics on the Workplace." *Personnel* 65 (September 1988), 26-31, 34, 36.

Covers advantages of robotics, societal changes, labor and labor institutions, the management, personnel, and structure of the corporation, some solutions to worker problems, and management's responsibility. Illustrated with photograph.

1437. Cusack, Michael. "Industrial Robots: How Are They Changing Our World?" *Senior Scholastic* 115 (February 4, 1983), 4-7.

Discusses how industrial robots are used, possible impact on jobs and human workers, and job opportunities. Includes photographs.

1438. Ebel, Karl-H. "The Impact of Industrial Robots on the World of Work." *International Labour Review* 125 (January/February 1986), 39-51.

Discusses the diffusion of industrial robots in the world, effects of robotization on employment, impact of industrial robots on working conditions, unions and industrial relations, and social and labor problems that may be created by robotics.

1439. Ettlie, John E.; Vossler, Marika L.; and Klein, Janice A. "Robotics Training." *Training and Development Journal* 42 (March 1988), 54-55.

Discusses studies of who is trained in companies using industrial robots and considers needs for training of various personnel. Illustrated with photograph.

1440. Fey, Carol. "Working with Robots: The Real Story." *Training* 23 (March 1986), 49-51, 56.

Discusses four realities: (1) robots are not replacing entire shifts of workers; (2) a robot is just a tool; (3) regular plant personnel maintain robots; (4) job category and seniority dictate who is trained. Describes a robot maintenance training program at the Ford Motor Company, Dearborn, Michigan. Illustrated with photograph.

1441. Foulkes, Fred K., and Hirsch, Jeffrey L. "People Make Robots Work." *Harvard Business Review* 62 (January/February 1984), 94-102.

Discusses the problem of introducing robots to the workplace. Considers factors in deciding to use robots, selecting a site, developing a worker displacement policy, gaining line management support, educating employees, working with the union, working with robots, and making employees comfortable. Illustrated with drawings.

1442. Globus, Paul. "Robots March on North American Industry." *Reader's Digest* (Canadian edition) 128 (March 1986), 83+.

1443. Guest, Robert H. *Robotics: The Human Dimension*. New York: Pergamon Press, 1984. 42 p.

Provides an introduction to industrial robot usage now and in the future, discusses the human and social effects of robotization, considers robots and the role of trade unions and governments, and gives guidelines for introducing robots in the workplace. Includes abstracts of related materials, recommended reading and additional reading lists.

1444. Guest, Robert H. "The Human Effects of Robots: Qualities of Life." *Vital Speeches of the Day* 49 (June 1, 1983), 510-513.

Discusses how robots will affect human beings at the workplace. Considers questions such as whether large numbers of workers will lose their jobs, job satisfaction, nature of work humans will perform in running robot systems, the issue of retraining, the role of trade unions, and role of government in bringing about adjustment to the coming age of robotization. Compares the United States and Japanese approaches and experiences.

1445. Hollon, Charles J., and Rogol, George N. "How Robotization Affects People." *Business Horizons* 28 (May/June 1985), 74-80.

Discusses results of a project that examined how major industrial concerns use robotics, impact of robotization on organizational and societal employment levels, responsibilities that industry assumes for retraining displaced labor personnel, worker reactions to robotization, and societal impact of robotization.

1446. Hopkins, Shirley A., and Hopkins, Willie E. "Robotics: A New Challenge for Human Resource Planning." *Human Resource Planning* 7 (1984), 197-202.

Covers current trends in robotics, implementations for human resource planning, and steps to take in a "Meeting the Robot Challenge" program.

1447. Howard, John M. "Focus on the Human Factors in Applying Robotics Systems." *Robotics Today* (December 1982). Reprinted in *Working Safely with Industrial Robots*, pp. 31-33. Edited by Peter M. Strubhar. Dearborn, MI: Robotics International, Society of Manufacturing Engineers, 1986. 266 p.

Considers work environment, worker-machine interface, job design, selection and training of personnel, maintainability, safety, management issues, and communications.

1448. Hunt, H. Allan, and Hunt, Timothy L. *Human Resource Implications of Robotics*. Kalamazoo, MI: W. E. Upjohn Institute for Employment Research, 1983. 215 p.
Examines the job creation and job displacement potential of industrial robots.

1449. Katzman, Marvin S. "When Robots Dominate the Workplace." Supervisory Management 28 (June 1983), 37-42.
Discusses displacement of workers and the deskilling of jobs and considers the effects. Illustrated with photograph.

1450. Keefe, Patricia. "Union Files Complaint Objecting to Robot." *Computerworld* 17 (November 28, 1983), 4.
Reports on union complaint charging that the United States Department of Labor's Mine Safety and Health Administration installed a robot to test mine dust samples, admittedly a "terrible and dirty job," replacing four nonunion workers (who came under the union's representation by law) without informing the union.

1451. Knod, Edward M., Jr.; Wall, Jerry L.; Daniels, John P.; Shane, Hugh M.; and Wernimont, Theodore A. "Robotics: Challenges for the Human Resources Manager." *Business Horizons* 27 (March/April 1984), 38-46.
Discusses ways to facilitate employee acceptance of robots into the workplace.

1452. Main, Jeremy. "Work Won't Be the Same Again." *Fortune* 105 (June 28, 1982), 58-61, 64-65.
Discusses the effect of computers and robots on workers in offices and factories and job retraining that is being done. Includes color photographs.

1453. Marshall, Harry. "The Robot Is Not an Enemy." *Modern Machine Shop* (March 1984), 36.

Reports on statements by Walter Wersel, president of the Robot Institute of America, concerning robots and factory workers.

1454. Martin, Gail M. "Industrial Robots Join the Work Force." *Occupational Outlook Quarterly* 26 (Fall 1982), 2-11.

Describes what industrial robots are, types of industrial robots, how they are used, their uses in the future, effects on the workforce, new occupations that may emerge as a result of industrial robot use, and retraining of workers. Includes photographs and drawings. Cover of this issue includes photograph of industrial robot.

1455. Milne, Emile. "The Robots Are Coming." *Black Enterprise* 12 (January 1982), 27-31.

Discusses the use of robots in manufacturing and implications for black workers. Illustrated with photographs.

1456. Minty, Gordon. "Robot Technicians: Is There a Need?" *Journal of Industrial Teacher Education* 24 (Spring 1987), 83-90.

Reports on results of a study to learn the technical specialties involved in maintaining, repairing, and operating robots in Michigan, advantages and disadvantages of using technical specialties that may cause a change in technical specialties used with robots in the future, and the prevalence and need of robot technicians in manufacturing plants in Michigan.

1457. Reed, David. "Don Williams: An American Worker's Comeback." *Reader's Digest* 128 (April 1986), 168, 171-173, 175.

Reports on how an unemployed pipe fitter in the steel industry retrained and obtained a job in the automotive industry maintaining spot-welding robots.

1458. Scarborough, Harry. "Working with Robots Is a Bore." *New Scientist* 90 (May 28, 1981), 555.
Discusses some of the problems workers have when working with robots and union reaction to robots. Illustrated with photographs.

1459. Shenkar, Oded. "Robotics: A Challenge for Occupational Psychology." *Journal of Occupational Psychology* 61 (March 1988), 103-112.
Considers human problems in robotization in the workplace such as blocked promotion, decrease in social interaction, and change in extrinsic and intrinsic rewards, and gives recommended intervention in robotized organizations, according to problem.

1460. Williams, Roger. "The Robot: Friend or Foe?" *Psychology Today* 17 (December 1983), 21.
Reports on a study of employee reactions to the introduction of a single robot at a manufacturing plant, conducted by Linda Argote, Paul S. Goodman, and David Schkade, of Carnegie Mellon University Graduate School of Industrial Administration, Pittsburgh, Pennsylvania.

1461. Zippo, Mary. "The Robot in Industry: Friend or Foe of Workers." *Personnel* 57 (November/December 1980), 51-52.
Discusses the impact of industrial robots on the workplace, considering worker replacement with robots.

SECTION 9. PHYSICAL IMPLICATIONS OF ROBOT USE

Section 9 concerns materials discussing or reporting on safety problems with robots, especially those used in the workplace. Robots can be dangerous and workers have been killed by them. Accidents often involve an industrial robot that starts moving again without warning while a worker is within the robot's work envelope, the area in which a robot, particularly a robotic arm, can work without obstruction.

Robots that can be dangerous have numerous safeguards such as automatic shut-off switches, fenced enclosures with gate and foot mat switches which are activated by human pressure, and sensors that shut off the robot if an unknown object is sensed. Many accidents occur because workers become careless and ignore or override safety mechanisms.

On the other hand, robots can prevent accidents by performing dangerous jobs that have previously been done only by human workers. Many examples of these uses can be found in Section 5, especially in the groupings under Other Robot Uses. Some of these uses include sentry duty in prisons and large warehouses, underwater inspection of cables, satellite repair in space, mining underground, work in nuclear power plants, bomb disarming, and hostage negotiation (one wall-climbing, window-washing robot so frightened a fugitive as it climbed into the window of the room in which he was hiding that he raced out into the arms of the police screaming).

Future robots may have capabilities that allow them to turn on humans and become killer robots. Materials throughout this sourcebook touch on that scenario. These materials can be located by consulting the Index under "killer robots."

Students will find materials to support both the pros and the cons of robot use in terms of physical implications in Section 9. Many materials will be particularly useful if used in combination with materials from other sections.

1462. "Looking Out for Number 7535." *Inc.* 7 (March 1985), 20.
Reports on use of small robot simulator by Advanced Analytics Corporation, Champaign, Illinois, to teach workers how to use larger, potentially dangerous robot. Illustrated with drawing.

1463. "Machines as Health Hazards." *Science News* 127 (April 6, 1985), 216.
Includes mention of first United States, robot-related death reported to National Institute for Occupational Safety and Health, which occurred in Michigan, in July 1984.

1464. "Robots." *National Safety and Health News* 132 (October 1985), 93-95.
A National Safety Council industrial data sheet on safe installation and operation of industrial robots. Illustrated with drawings.

1465. Altamuro, Vincent M. "Working Safely with the Iron Collar Worker." *National Safety News* (July 1983), 38-40. Reprinted in *Working Safely with Industrial Robots*, pp. 73-75. Edited by Peter M. Strubhar. Dearborn, MI: Robotics International, Society for Manufacturing Engineers, 1986. 266 p.
Discusses fatal accidents involving industrial robots, main reasons for these accidents, and safety measures to take to prevent them.

1466. Bonney, Maurice C., and Yong, Y. F., eds. *Robot Safety*. Bedford, England & Berlin: IFS (Publications) Ltd. & Springer-Verlag, 1985. 295 p.
Covers papers dealing with robot safety, under the following topics: legislation and standards; surveys and analyses; safety system design; implementation; methodology;

safety system components. Ends with a section on case studies. Includes illustrations and photographs.

1467. Carrico, Lloyd R. "Three Keys to Robot Safety: Training, Design, and Implementation." *National Safety News* 129 (May 1984), 81-85.

Describes ways to design robotic applications for safety and discusses training and implementation. Illustrated with photographs.

1468. Cox, John L., and Butler, J. Kent. "Human Factors Issues in Robotics: Physical, Mental, Safety, Legal." In *Working Safely with Industrial Robots*, pp. 34-45. Edited by Peter M. Strubhar. Dearborn, MI: Robotics International, Society of Manufacturing Engineers, 1986. 266 p.

Covers physical, mental, safety, and legal issues to be considered when using industrial robots.

1469. Diekemper, Roman F. "Facing the Exposures of the Robot Revolution." *Risk Management* 31 (October 1984), 40-42, 44, 46.

Covers protecting employees and bystanders, preventing contact between dangerous parts and workers, potential for physical damage, financial losses from damaged or disabled robots, managing hazards, and product liability considerations. Illustrated with drawing.

1470. Freifeld, Karen. "In the Lion's Cage." *Forbes* 136 (October 7, 1985), 142, 144.

Reports on industrial robot-related accidents and ways to make industrial robots safer.

1471. Fuller, John G. "Death by Robot." *Omni* 6 (March 1984), 44-46, 97-98, 100, 102.

Provides detailed account of events leading up to and following death of Bob Williams by robot at Ford Motors' Michigan Casting Center, Flat Rock, Michigan. Illustrated with painting.

1472. Gillin, Paul. "Kin of Man Killed by Robot Awarded $10 Million." *Computerworld* 17 (August 15, 1983), 8.
Reports on jury award to family of worker struck in the head and killed by an industrial robot while retrieving a part.

1473. Lambrinos, James, and Johnson, William G. "Robots to Reduce the High Cost of Illness and Injury." *Harvard Business Review* 62 (May/June 1984), 24, 26, 28.
Discusses the use of robots to reduce occupational accidents in manufacturing and to remove workers from hazardous environments.

1474. Lauck, Kenneth E. "Safeguarding Industrial Robots." *National Safety News* 129 (April 1984), 63-67.
Describes ways to safeguard industrial robot installations to prevent accidents, based on the General Motors Safety Standard, and includes a listing of the Standard.

1475. Lodge, James E. "How to Protect Robot Maintenance Workers." *National Safety News* 129 (June 1984), 48. Reprinted in *Working Safely with Industrial Robots*, pp. 178-181. Edited by Peter M. Strubhar. Dearborn, MI: Robotics International, Society of Manufacturing Engineers, 1986. 266 p.
Considers hazard types and sources and ways to prevent accidents. Includes photographs and drawing.

1476. Minter, Stephen G. "High Tech: Its Implications for Safety Management." *Occupational Hazards* 46 (December 1984), 47-51.
Includes a discussion on industrial robot safety problems. Provides color photograph.

1477. Morgan, Doris J. "Safety Considerations in Robot Welding Operations." *National Safety News* (July 1984). Reprinted in *Working Safely with Industrial Robots*, pp. 82-86. Edited by Peter M. Strubhar. Dearborn, MI: Robotics

International, Society of Manufacturing Engineers, 1986, 286 p.

Discusses codes and standards, layout and facilities, guarding, controls, procedures, training, and plant requirements.

1478. Peterson, Ivars. "Stepping into Danger." *Science News* 130 (July 12, 1986), 28-31.

Reports on uses of robots in hazardous occupations, such as mining, construction, gas leak repair, underground pipeline and bridge inspection, and nuclear power plant work. Includes photographs. Cover of this issue shows a robot in a mine shaft.

1479. Strubhar, Peter M., ed. *Working Safely with Industrial Robots*. Dearborn, MI: Robotics International, Society of Manufacturing Engineers, 1986. 266 p.

Covers operation of industrial robots, human factors, philosophies of robot safety, sophisticated safety, installation, maintenance, construction of safety devices, safety training, and safety standards. Includes photographs and drawings.

1480. van Deest, Robert. "Robotic Safety - a Potential Crisis." *Best's Review* 85 (August 1984), 74-77, 95.

Discusses safety problems of industrial robots, lists causes of robot-caused injuries, with percentages, and describes ways to lower the accident rate through training programs and safety techniques and devices.

1481. Wilcher, Frank E., Jr. "High Technology Safety: Meeting Tomorrow's Challenges Today." *Occupational Hazards* 46 (October 1984), 21-22, 25.

Includes discussion on industrial robot safety and safety procedures to follow. Provides photograph.

1482. Ziskovsky, Joseph P. "Working Safely with Industrial Robots." *Plant Engineering* 38 (May 10, 1984), 81-85.

Discusses the robot environment, types of hazards, physical safeguards, and safe work practices, and lists procedures for entering a robot work area. Illustrated with photograph and drawings.

SECTION 10. NONPRINT MEDIA

Included in this section are items that could be classified as audiovisual materials as well as computer programs. The section also includes a few computer program reviews. Nonprint materials are often listed in special catalogs dealing with a particular format. Films and videorecordings may be bought, rented, or borrowed from film and videorecording libraries and distributors. Directories covering nonprint formats should be checked for sources and for new items. A list of nonprint media producers and distributors for the materials listed in this section can be found in Appendix F.

COMPUTER PROGRAMS

General Information

1483. *Experiments in Artificial Intelligence: Robotics*. Pleasantville, NY: Human Relations Media. For Apple II series, Commodore.

Students experiment with artificial intelligence concepts by designing and controlling simple machines and small household appliances, using a personal computer. A robot is used to operate devices such as switches, coffee pots, and lamps.

1484. *History of the Robot*. Distributed by Hearlihy & Company. For Apple II series.

Tutorial computer-assisted instruction. Covers early animation, robots in entertainment, and robots of the future.

1485. *Robot Odyssey*. Menlo Park, CA: The Learning Company. For Apple II series.

Educational game in which player designs, builds, and programs robots. Robots must be programmed correctly in order to escape from an underground maze. Includes operating manual with game strategies. See number 1487 for information on playing *Robot Odyssey*.

1486. Costello, Matthew J. "On Gaming." *Analog Science Fiction & Science Fact* 107 (September 1987), 81, 128.

Describes *Robotic Workshop* (Multibotics, Incorporated, Access Software, Incorporated, 2561 South 1560 West, Suite A, Wood Cross, UT 84087), a game to make, test, and play with computer-directed robots, which comes with connectors, motors, and other equipment to create robots. Available for a number of microcomputers.

1487. Dewdney, A. K. "Computer Recreations; A Circuitous Odyssey from Robotropolis to the Electronic Gates of Silicon Valley." *Scientific American* 253 (July 1985), 17, 19.

First part of column describes the educational game *Robot Odyssey*, and the author's experiences in playing it. Illustrated with color and noncolor drawings and photographs.

1488. Mace, Scott. "Logic Games: Tools for Thought." *Personal Computing* 9 (March 1985), 265.

Reviews *Robot Odyssey I*. Illustrated with drawing. See number 1485.

1489. Rappaport, Susan. "Software to Teach Problem Solving." *Library Journal* 110 (May 15, 1985), 44-45.

Includes information on *Robot Odyssey*. See number 1485.

Secondary Grades Level

1490. *Robotics and Computers*. Distributed by Career Aids and Hearlihy & Company. For Apple II series, IBM PC with 128K.

Beginning level. Tutorial computer-assisted instruction. Six lessons include: (1) What is a Robot?; (2) Robot Anatomy and Movement; (3) Basic Robot Programming; (4) Robot Applications; (5) The Decision to Implement; (6) Robots and Your Future. Includes manual.

1490a. *Robotics Issues and Dilemmas*. Distributed by Hearlihy & Company. For Apple II series.

Tutorial computer-assisted instruction. Three lessons include: (1) Robot Pros and Cons; (2) Robot, the Job Stealer?; (3) Artificial Intelligence in Robotics.

Industrial Robots

1491. *Industrial Robots*. Distributed by Hearlihy & Company. For Apple II series, IBM PC, IBM PS/2.

Tutorial computer-assisted instruction. Covers what an industrial robot is, the automated factory, and robots that handle materials in palletizing and die casting.

1492. *Robot Assembly Lab I*. Charlottetown, PEI, Canada: Sigler Software, Inc. Available from Sigler Software and Hearlihy & Company. For Apple II series.

Beginning level. Simulation of a robot design laboratory. Allows student to create custom robots that are then evaluated by the program and redesigned or modified by the student.

1493. *Robotics Fundamentals*. Distributed by Hearlihy & Company. For Apple II series, IBM PC with 128K.

Beginning level. Tutorial computer-assisted instruction. Covers robotics parts identification and terminology, industrial robots, microcomputer controllers, robot control programming, and robot sensing systems. Includes five reproducible quizzes and answer keys.

1494. *RobCat.* Distributed by Career Aids, Inc. and Hearlihy & Company. For Apple II series, TRS-80 Model III/4, Commodore 64/128.
Advanced level. Industrial robot simulation on robot programming. Covers four application areas: (1) palletizing; (2) machine loading and unloading; (3) visual inspection; (4) spot welding. Use with Robcat 4-axis robot. Includes instruction manual.

Robot Characters Used in Teaching

1495. *ChipWits.* Sunnyvale, CA: Epyx (Commodore version); Calabasas, CA: Brainworks, Inc. (Macintosh version). For Apple II series, Commodore 64, Macintosh.
An educational game whose objective is to teach the fundamentals of programming. For an extensive review see number 1496.

1496. Anderson, John J. "ChipWits: Bet You Can't Build Just One." *Creative Computing* 11 (December 1985), 76, 78-79.
Reviews the computer robot game *ChipWits*, whose objective is to teach systematic thinking. Illustrated with drawings and a screen display. Cover of this issue includes picture of the robot Armatron.

1497. *Karel the Robot.* Distributed by The Center for Humanities, Inc.
Teaches structured programming and modularization in UCSD PASCAL using a robot simulator which, when programmed, performs different tasks.

1498. *Logo Robot.* Distributed by Scholastic Inc. For Apple II series, Commodore 64, IBM PC/PCjr.

Grades 4 through 8. Tutorial computer-assisted instruction on the Logo programming language. Includes student handbook.

1499. *Macrobots*. Las Vegas, NV: Unicorn Educational Software. For Macintosh series.
Preschool level. Nursery tale introducing concepts of counting, addition, subtraction, and letter recognition. Includes robot construction kit.

1500. *Robomath*. Distributed by K-12 MicroMedia. For Apple II series, IBM PC.
Grade 2 and up. Drill and practice computer-assisted instruction on multiplication and division. Uses robo-mathematicians.

1501. *Robot Probe*. Distributed by Sunburst Communications. For TRS-80 III, 4.
Grade 3 to adult. Educational game to teach computer programming, using a robot that must be programmed. Includes activity cards, reproducible programming sheets, teacher's guide.

1502. *Robot Writer*. Random House Media. For Apple II series.
Grade 1 and up. "Creative Writing Series." Writing program in which user chooses robot parts from body shop, creates a robot, writes about it, and prints out a robot report.

1503. *Ten Little Robots*. Las Vegas, NV: Unicorn Software. For Apple II series, Atari/Commodore 64, IBM PC/PCjr.
Preschool through grade 2. Educational games using robots introduce counting, addition, subtraction, and recognition of letters. A fifth game is for doodling and creative activity.

SOUND RECORDINGS

Classroom Materials

Elementary Grades Level - Fiction

1504. Bunting, Eve. *The Robot People*. Instructional Aids, 1978. 1 cassette, 2-track. Also available in book format. See number 546.

An adventure story about two robots, Link and Magnus, who work in an underground nuclear operation, and what happens when the son of a scientist working there learns that Magnus has developed a dislike of all humans.

1505. Fettig, Art. *The Three Robots*. Nightingale-Conant Corporation; Growth Unlimited, Inc., 1981. 1 cassette. Also available in book format. See number 556.

The story of three robots, one happy, one neutral, and one sad, and how all became happy robots with positive attitudes.

1506. Fettig, Art. *The Three Robots and the Sandstorm*. Battle Creek, MI: Growth Unlimited, Inc., 1981. 1 cassette. Also available in book format. See number 557.

Continues the story of the three robots, illustrating the teamwork it takes to overcome handicaps.

1507. Fettig, Art. *The Three Robots Discover Their Pos-Abilities*. Battle Creek, MI: Growth Unlimited, Inc. 1 cassette. Also available in book format. See number 558.

Continues the adventures of the three robots, illustrating the value of goal setting.

1508. Fettig, Art. *The Three Robots Find a Grandpa*. Battle Creek, MI: Growth Unlimited. 1 cassette. Also available in book format. See number 559.

Continues the story of the three robots, illustrating a lesson in love.

Secondary Grades Level - Fiction

1508a *Isaac Asimov's Robot City. Odyssey.* By Michael P. Kube-McDowell. New York: Caedmon, 1988. 1 cassette. Also available in book format. See number 611.

SLIDES

General Information

1509. *Robotics Explained*. Uniondale, NY: Bergwall Productions, 1983.
Shows the basic functions and movements of robots.

Industrial Robots

1510. *Computer Applications. 10. Robotics*. Mill Valley, CA: Prismatron Productions, Inc., 1985. 55-60 slides, 1 cassette, teacher's guide. Also available on filmstrip. See number 1522.
Covers industrial robots, uses, control, and sensors.

FILMSTRIPS

General Information

1511. *From Fiction to Fact*. Pleasantville, NY: Educational AudioVisual, Inc. 1 filmstrip, 1 cassette. Teacher's guide.
Set with *Robots in Your Future*. Covers types of robots, mechanical sight and sensory devices, and how robots can emulate some basic human activities.

1512. *Frontiers of Technology*. Washington: National Geographic Society, 1985. 3 filmstrips, color, 3 cassettes, 4 guides.

1513. *Robotics*. Tarrytown, NY: Prentice-Hall Media, 1983. 1 filmstrip, color, 1 cassette, program guide.

1514. *Robotics*. Mill Valley, CA: Prismatron Productions, 1980. 1 filmstrip, color, 1 cassette, booklet.

1515. *Robotics*. Chicago: Society for Visual Education, Inc., 1984. 2 filmstrips, 2 cassettes, teacher's manual, 16 activity sheet masters.

Discusses history of robots, relationship of robots and computers, artificial intelligence, and impact of robots on employment. See specific information under *Robots at Work*, number 1517, and *Robots in Our World*, number 1518.

1516. *Robotics and Artificial Intelligence*. Westminster, MD: Random House, 1984. Distributed by Career Aids, Inc., and others. 1 filmstrip, color, 1 cassette, guide.

Shows what robots are, how they are used, and how they work, including how computer chips are combined with programmed instructions and sensing devices to enable robots to perform humanlike functions. Tells how artificial intelligence sets robots apart from other machines.

1517. *Robots at Work*. Chicago: Society for Visual Education, Inc., 1984. 1 filmstrip, 1 cassette, activity sheet masters.

Shows how robots are used in industry, space exploration, and research, considers the personal robot, and discusses the impact that robots are likely to make on our future. May be bought as a set with *Robots in Our World*, and with or without skill sheets. Skill sheets may be bought separately. See general note at number 1515.

1518. *Robots in Our World*. Chicago: Society for Visual Education, Inc., 1984. 1 filmstrip, 1 cassette, activity sheet masters.

Surveys history of the development of robots, basics of robot operation, what robots can and cannot do, how they are designed, and how they "learn" to perform various tasks. May be bought as a set with *Robots at Work*, and with or without skill sheets. Skill sheets may be bought separately. See general note at number 1515.

1519. *Robots in Your Future*. Pleasantville, NY: Educational AudioVisual, Inc. 1 filmstrip, 1 cassette. Teacher's guide.
Set with *From Fiction to Fact*. Covers types of robots, mechanical sight and sensory devices, and how robots can emulate some basic human activities.

1520. *The Mystery of Man-Made Man*. Westminster, MD: Random House School Division. 1 filmstrip, 1 cassette.
Considers questions such as "Can we make robots that are able to do just about anything - even outthink us?"

Elementary Grades Level

1521. *All About Robots*. Stamford, CT: Educational Dimensions Group, 1984. 2 filmstrips, 41 and 45 frames, color.
Includes *Meet the Robots* and *Robots at Work*. Shows what robots are, their capabilities, and applications.

Industrial Robots

1522. *Computer Applications. 10. Robotics*. Mill Valley, CA: Prismatron Productions, Inc., 1985. 55-60 frames, 1 cassette, teacher's guide. Also available on slides. See number 1510.
Covers industrial robots, uses, control, and sensors.

1523. *Robotics*. 1985. Distributed by Career Aids, Inc., separately or as a set, and by Library Filmstrip Center as a set. 6 filmstrips, color, 6 cassettes, 6 teacher's guides. Also available on videocassette. See number 1544.
Titles include: (1) Robots in Industry; (2) Robot Sub-Systems; (3) Operating Parameters of Robots; (4)

Applications of Robot Technology; (5) Implementation of Robots; (6) Human Factors in Robotics.

1524. *Robotics Explained*. Uniondale, NY: Bergwall Productions, 1983. Distributed by Career Aids, Inc. 4 filmstrips, color, 4 cassettes, study guide. Also available on videocassette. See number 1568.

Filmstrips in the series are: (1) Robot Nomenclature; (2) Robots in the Workplace; (3) Robot Classification; (4) Basics of the TEACHMOVER Robot.

1525. *Robots, Selection and Installation*. By James W. Hannemann. Garden City, NY: Bergwall Productions, 1985. 4 filmstrips, color, 4 cassettes.

Titles include: (1) Robotic Design; (2) Principles of Selection; (3) Assembling a Robot; (4) Installing and Programming.

FILMS

General Information

1526. *Bionics: Man or Machine?*. Avatar Learning, Inc., 1977. Distributed by Encyclopaedia Britannica Educational Corporation. 24 minutes., color, 16 mm. Also available on videocassette. See number 1539.

Discusses bionics - the science that copies nature. Robots that replace people on jobs, robots that make decisions, and machinery that extend our sight, hearing, and mobility, or replace anatomical parts, bring about the question, "Are we masters of our own inventions?"

1527. *Miraculous Machines*. Washington: National Geographic Society, 1985. 59 minutes, color, 16 mm. Also available on videocassette, under the titles *Miraculous Machines* (see number 1542), and *The Miracle Chip* (see number 1550).

Includes examples of robots used with the handicapped, with older persons, in Japan (Marilyn Monroe singing, robot handing out papers, robot playing the organ), and in industry.

1528. *Robotics: The Future Is Now*. Van Nuys, CA: AIMS Media, 1984. 20 minutes, color, 16 mm. Also available on videocassette. See number 1545.

Shows applications of robots, particularly in industry, and how they may be used in the future.

1529. *Robots: The Computer at Work*. Produced by Gakken Films. Distributed by AIMS Media. 22 minutes, color, 16 mm., teacher's guide. Also available on videocassette. See number 1548.

Shows robot applications.

1530. *The Age of Intelligent Machines*. By Raymond Kurzweil. Waltham, MA: The Kurzweil Foundation, 1987. Distributed by AIMS Media. 29 minutes, color, 16 mm. Also available on videocassette. See number 1549.

Covers recent developments in artificial intelligence and applications of robots, including machines designed to aid the handicapped.

1531. *The Robot Revolution*. Chicago: Encyclopaedia Britannica Educational Corporation, 1984. 19 minutes, color, 16 mm.

Surveys robot capabilities and limitations, potential for improving the quality of life, and implications for the labor force.

1532. *The Robotic Revolution*. Washington: National Geographic Society, 1986. 24 minutes, color, 16 mm., guide. Also available on videocassette. See number 1554.

Shows how robots are used in Japan in the automobile, electronic, merchandising, and other industries.

Recent History

1533. *Robotics: Isaac Asimov's (Clank! Clank!) Artificial Man*. New York: Cinema Guild, 1973. Distributed by Hearst Metrotone News and Document Associates, Inc. 22 minutes, color, 16 mm.

Isaac Asimov comments on developments in the field of robotics. Shows General Electric's Hardiman and Quadruped, Unimation's Ultimate, and research robots at Johns Hopkins University, Baltimore, Maryland, and Columbia University in New York City.

1534. *Robots Get Smarter*. Berkeley: KCET; University of California Extension Media Center, 1970. 29 minutes, color, 16 mm.

Demonstrates robots being developed at the Stanford Research Institute, Stanford, California: Versatran, a robot that uses hand tools in dangerous environments; Butterfinger, a robot with an arm, eye, brain, and limited coordination; and Shakey, a mobile robot.

Industrial Robots

1535. *Making the Quality Outboard*. Outboard Motor Corporation, 1984. Distributed by Solana Studios. 26 minutes, color, 16 mm.

Shows how Evinrude and Johnson outboard motors are made, and how robots, as well as other advanced equipment, are used in the process.

1536. *Robots: Intelligent Machines Serving Men*. New York: Allegro Film Productions, Inc., 1981. 14 minutes, color, 16 mm.

Shows how robots are taking over jobs, producing safer working conditions, higher quality products, and more efficient use of resources. Considers robots with limited "touch," "sight," and "judgment."

1537. *The Robots Are Coming.* London: BBC Enterprises Film and Video Library. Color, 16 mm.

Explores industrial applications, and the implications of Japan's greater number of robots compared to Great Britain.

Other Applications

1538. *Ballet Robotique.* Santa Monica, CA: Pyramid Film & Video, 1982. 8 minutes, color, 16 mm. Also available on videocassette. See number 1578.

A montage of film of industrial robots at work, set to music to suggest a choreographed performance.

VIDEO RECORDINGS

General Information

1539. *Bionics: Man or Machine?.* Avatar Learning, Inc., 1977. Distributed by Encyclopaedia Britannica Educational Corporation. 1 cassette, 24 minutes, color, U Standard 3/4 in., Beta 1/2 in., VHS 1/2 in. Also available on film. See number 1526.

Discusses bionics - the science that copies nature. Robots that replace people on jobs, robots that make decisions, and machinery that extends our sight, hearing, and mobility or replaces anatomical parts bring about the question, "Are we masters of our own inventions?"

1540. *CNN Special Report: Me and My Robot,* Mt. Kisco, NY: The Center for the Humanities, 1988. Distributed by Guidance Associates. 1 cassette, 19:55 minutes, VHS, 1/2 in., 3/4 in., teacher's guide.

Shows uses of robots, where they can be found, and future impact.

1541. *Decisions, Decisions, Decisions*. New Jersey Network, 1983. Distributed by Agency for Instructional Technology. 1 cassette, 20 minutes, color, teacher's guide.

Program 2 of *You, Me, and Technology* series. Discusses information processing and shows that the steps in decision making are the same whether the input is processed by a machine such as a robot, an electronic amplifier, or a human brain. Considers artificial intelligence and the role of human value judgments.

1542. *Miraculous Machines*. Now titled *The Miracle Chip*. Washington: National Geographic Society, 1985. 1 cassette, 59 minutes, color, VHS, Beta, U-Matic 3/4 in. Also available on film. See number 1527.

Includes examples of robots used with the handicapped, with older persons, in Japan (Marilyn Monroe singing, robot handing out papers, robot playing the organ), and in industry.

1543. *Robotics*. San Mateo, CA: KCSM-TV, 1983. 1 cassette, 22 minutes, color, 3/4 in.

Discusses what a robot is and what its roles may be in the future.

1544. *Robotics*. 1985. Distributed by Career Aids, Inc. 1 cassette, VHS. Also available on 6 filmstrips. See number 1523.

Covers the following topics: (1) Robots in Industry; (2) Robot Sub-Systems; (3) Operating Parameters of Robots; (4) Applications of Robot Technology; (5) Implementation of Robots; (6) Human Factors in Robotics.

1545. *Robotics: The Future Is Now*. Van Nuys, CA: AIMS Media, 1984. 1 cassette, 20 minutes, color. Also available on film. See number 1528.

Shows applications of robots, particularly in industry, and how they may be used in the future.

1546. *Robotics Revolution*. Chicago: Society for Visual Education. 1 cassette, 27 minutes.

Covers history, robot operation, what robots can and cannot do, how robots are designed, how robots are "taught," and uses in the future.

1547. *Robots*. By Charles Rosen. Sacramento: California State University, Center for Instructional Media, 1982. 2 cassettes, 90 minutes, color, 3/4 in.

1548. *Robots: The Computer at Work*. Produced by Gakken Films. Distributed by AIMS Media. 1 cassette, 22 minutes, color, teacher's guide. Also available on film. See number 1529.

Shows robot applications.

1549. *The Age of Intelligent Machines*. By Raymond Kurzweil. Waltham, MA: The Kurzweil Foundation, 1987. Distributed by AIMS Media. 1 cassette, 29 minutes, color, VHS, 3/4 in. U-Matic. Also available on film. See number 1530.

Covers recent developments in artificial intelligence and applications of robots, including machines designed to aid the handicapped.

1550. *The Miracle Chip*. Formerly titled *Miraculous Machines*. Washington: National Geographic Society. 1 cassette, 59 minutes, color, VHS. Also available on videocassette with the title *Miraculous Machines* (see number 1542), and on film as *Miraculous Machines* (see number 1527).

Includes examples of robots used with the handicapped, with older persons, in Japan (Marilyn Monroe singing, robot handing out papers, robot playing the organ), and in industry.

1551. *The Robot Revolution*. Deerfield, IL: Coronet Films, 1985. 1 cassette, 58 minutes, color.

Nova series. Shows how industry is using computers and robots to increase productivity and examines the future workplace and the role of robots and technology in changing the way people work.

1552. *The Robot Revolution*. Chicago: Encyclopaedia Britannica Educational Corporation. 1 cassette, 19 minutes, color, U-Matic, teacher's guide.

Examines the myth and the magic of robots and looks at some of the ways we will think about and use them in the future.

1553. *The Robot Revolution*. 1984. Distributed by Journal Films, Inc. 1 cassette, 26 minutes, color, 1/2 in., 3/4 in.

Explores history of robots, how robots are used, how they might be used in the future, and how they have been accepted.

1554. *The Robotic Revolution*. Washington: National Geographic Society, 1986. 1 cassette, 24 minutes, color, guide. Also available on film. See number 1532.

Shows how robots are used in Japan in the automobile, electronic, merchandising, and other industries.

1555. *The Robots Are Coming*. London: BBC Enterprises, 1979. 1 cassette, 50 minutes, color, 3/4 in.

Robot Features

1556. *Does That Compute?* Evanston, IL: Journal Films, 1984. 1 cassette, 26 minutes, color.

Demonstrates differences between human and robotic perception and explains how robots respond to vocal commands.

1557. *Introducing the Robot*. Wilmette, IL: Films Incorporated. A BBC-TV production. 1 cassette, 25 minutes, color, VHS.

Program 1 of *Computers in Control* series. Demonstrates how robots are controlled, considering sensors that detect signals, activators, and important computer software.

1558. *Robots That Run*. By Marc Raibert. Cambridge: The MIT Press. 1 cassette, 15 minutes, color, VHS, other formats by special order.

Prepared to accompany *Legged Robots That Balance* by Marc Raibert (see number 236). Provides historical overview and chronicles the development and behavior of four robots designed by Raibert and his co-workers at Carnegie Mellon University, Pittsburgh, Pennsylvania.

Career Information

1559. *Career Explorations: Robotics Technician*. Distributed by Charles Clark. 1 cassette, 5 minutes, VHS, Beta.

Profiles a robotics technician, tours the person's workplace, and explains the practical skills and training offered by employers and by colleges.

1560. *World of Work*. Tecnivision, Inc., 1984. Distributed by GPN. 2 cassettes, 60 minutes each, or 3 laser videodiscs.

Program 1, *Is Technology Stealing Your Job?*, provides an overview of computers, communications, and high technology fields such as robotics, considers their impacts on the workforce, and forecasts future requirements for the world of work. Program 2, *Where Will You Work Tomorrow?*, explores high-technology job opportunities, gives guidance regarding training opportunities, and shows how individuals can learn techniques to cope with technology's impacts on their jobs and futures. See *World of Work*, number 1586, for information on Program 3, *Guide to Tomorrow's Jobs*.

Industrial Robots

1561. *Computer Applications: Manufacturing and Design*. Chicago: Encyclopaedia Britannica Educational Corporation, 1982. 1 cassette, 27 minutes, color.

Covers the use of computer graphics to make models for product design or decision making. Includes examples of combining robotics with graphics in product manufacture.

1562. *Factory of the Future*. Distributed by Career Aids, Inc. 1 cassette, 62 minutes, VHS, 1/2 in.

Shows how a factory operates and how computers and robots can be used.

1563. *Fundamentals of Industrial Robotics*. Bloomington, IL: Meridian Education Corporation, 1986. Distributed by Career Aids, Inc.; Hearlihy & Company; and Library Filmstrip Center. 1 cassette, 25 minutes, color, VHS, Beta I, Beta II, U-Matic 3/4 in., teacher's guide. Also available in interactive video format. See number 1587.

Covers industrial robot terminology, fundamental concepts, limitations, programming, components and classification, industrial applications, and careers in robotics.

1564. *Justification, Applications, and Predictions in Robotics*. Dearborn, MI: Society of Manufacturing Engineers, 1983. 1 cassette, 55 minutes, color, VHS.

Panel of six industrial robotics experts discusses new methods, justification for robot use, and application ideas for the future.

1565. *Robot Safety at Caterpillar*. Dearborn, MI: Society of Manufacturing Engineers. 1 cassette, 12 minutes, color.

Produced by Caterpillar Tractor Company. Shows a variety of potentially hazardous situations involving robots and examines devices and procedures to avoid them.

1566. *Robotics*. Fullerton, CA: Fullerton College, 1984. 1 cassette, 30 minutes, color, VHS.

Shows computer-assisted manufacturing equipment and tells how it affects goods production. Explains that U. S. Amada, Limited, based in Japan, is bringing to the United States the latest in numerically controlled machines and robots.

1567. *Robotics: How It Is Changing the Workplace*. Distributed by Johnson/Rudolph Educational Resources. 1 cassette, 14 minutes, VHS, teacher's guide.

Associated Press Special Report. Shows robot applications and social implications and how negative effects of automation can be minimized through job retraining.

1568. *Robotics Explained*. Uniondale, NY: Bergwall Productions, 1983. Distributed by Career Aids, Inc. 4 cassettes, 47 minutes, color, VHS, 1/2 in., 3/4 in. Also available on filmstrips. See number 1524.

Titles are: (1) Robot Nomenclature; (2) Robots in the Workplace; (3) Robot Classification; (4) Basics of the TEACHMOVER Robot.

1569. *Robotics Technology: Beyond Basics*. Distributed by Career Aids, Inc. 6 cassettes, VHS, program guide.

Titles in this series are: (1) Robotic Evolution; (2) Robot Safety; (3) Robot Classification and Components; (4) Robot Programming; (5) Maintenance; (6) Robot Applications.

1570. *Robots 6: Tomorrow's Technology on Display*. Dearborn, MI: Society of Manufacturing Engineers. 1 cassette, 47 minutes, color, U-Matic.

Produced at the Robots 6 Exposition in 1982. Shows robots of all capabilities and sizes and includes an interview featuring two robotics industry pioneers, Joseph Engelberger and George Devol.

1571. *Robots 7: The Evolution in Automation*. Dearborn, MI: Society of Manufacturing Engineers. 1 cassette, 30 minutes, color.

Produced at the Robots 7 Exposition in 1983. Features models from robot manufacturers around the world performing various applications such as material handling, welding, spraying, and palletizing. Covers advances in robotic technology, tactile sensors, and machine vision.

1572. *Robots 8: Steel Collar Workers of the 80's*. Dearborn, MI: Society of Manufacturing Engineers. 1 cassette, 35 minutes, color.

Produced at Robots 8 Exposition in 1984. Shows material handling, welding, painting, assembly, water jet cutting, and non-contact gaging applications. Also includes lasers, automated guided vehicles, and computer-aided design systems, and ways they can be integrated with robotic systems.

1573. *Robots in Assembly and Packaging*. Dearborn, MI: Society of Manufacturing Engineers, 1985. 1 cassette, 46 minutes, color, VHS.

Five in-depth case studies show industrial robots in action. Applications include assembly of telephones, heating and air conditioner controls, faucet components, and terminal connectors, packaging of faucet components, and vendor pre-packaging for ease of robot use at International Business Machines (IBM).

1574. *Robots in Welding and Painting*. Dearborn, MI: Society of Manufacturing Engineers, 1985. 1 cassette, 35 minutes, color.

Five in-depth case studies show industrial robots in action. Applications include spot welding, arc welding, painting automobiles, and painting antennas.

1575. *Steel Collar Workers of the 80's*. Dearborn, MI: Society of Manufacturing Engineers. 1 cassette, 36 minutes.

Provides facts necessary to help planner determine the role of robots in future applications, and a picture of the current level of technology in automated manufacturing equipment.

1576. *The Marvels of Industrial Robots*. Distributed by Coronet Films & Video. 1 cassette, 36 minutes.
Shows industrial robot applications.

1577. *The Robot Revolution*. New York: Hearst Metrotone News, 1981. 1 cassette, 14 minutes, color, Beta, VHS, 3/4 U-Matic.
Shows how robots are transforming factory operations and production output around the world.

Other Applications

1578. *Ballet Robotique*. Santa Monica, CA: Pyramid Film & Video, 1982. 1 cassette, 8 minutes, color. Also available on film. See number 1538.
A montage of film of industrial robots at work, set to music to suggest a choreographed performance.

1579. *Robots in the Air and Space: The Sky Is Not the Limit*. By Stanley Weiss. Dearborn, MI: Society for Manufacturing Engineers, 1987.

1580. *Robots in the Service Industries*. By Joseph F. Engelberger. Dearborn, MI: Society for Manufacturing Engineers, 1987.

Classroom Uses

1580a. *Automation and Control Training Tape*. Modesto, CA: fischer america, inc., 1990. 1 cassette, 120 minutes. Available from fischer america, inc., 2420 Vanlayden Way, Modesto, CA 95356.

Teacher-training video on how to use fischertechnik robots in the classroom, and how to control analog and digital, input/output devices with the computer, using fischertechnik kits.

1580b. *Robot Model Kit.* Modesto, CA: fischer america, inc., 1989. 1 cassette, 14 minutes. Available from fischer america, inc., 2420 Vanlayden Way, Modesto, CA 95356.

Tells about fischertechnik robot kits (product overview) and covers basic difference between fixed sequence and variable sequence robots.

Classroom Materials

Fiction

1581. *C. L. U. T. Z.* New York: CBS/FOX, Inc., 1989. 1 cassette, 25 minutes, color, VHS.

Based on story of the same name by Marilyn Z. Wilkes. Tells the story of C. L. U. T. Z., a robot hired by a family to do their domestic chores, and the adventures he had while working there. See numbers 591, 592.

1582. *Robbie the Rascal.* Hollywood, CA: Paramount Home Video, 1985. Distributed by Media Basics Video. 1 cassette, 90 minutes, color, VHS.

Animated cartoon about good and evil, using a "cybot."

1583. *Robotman and Friends.* UFS, Inc. Distributed by Children's Video Library. 1 cassette, 48 minutes, color, VHS, Beta, U-Matic 3/4 in.

Shows the adventures of a robot and his friends who fight the forces of evil, in animated format.

1584. *Robotman & Friends II: I Want to Be Your Robotman.* Stamford, CT: Children's Video Library. 1 cassette, 42 minutes, color.

Shows the adventures of a robot and his friends and their fight against evil denizens.

1585. *The Black Hole*. Burbank, CA: Walt Disney Productions, 1979. Distributed by The Video Schoolhouse. 97 minutes, color, Beta, 1/2 in. Also available on LaserVision videodisc.

Tells the story of the crew of the spaceship, U. S. S. *Palomino*, including the robot, V.I.N.C.E.N.T., and their encounter with the human captain, and humanoid and robot crew of the U. S. S. *Cygnus*, near the Black Hole.

INTERACTIVE VIDEO

Career Information

1586. *World of Work*. Tecnivision, Inc., 1984. Distributed by GPN. Interactive videodisc.

Program 3, *Guide to Tomorrow's Jobs,* presents a visual database of information about different high technologies, job trends, skill requirements, and training sources. See number 1560 for information on Programs 1 and 2 of this series.

Industrial Robots

1587. *Fundamentals of Industrial Robotics*. Laser videodisc. Distributed by Library Filmstrip Center. Also available on videocassette. See number 1563.

Covers industrial robot terminology, fundamental concepts, limitations, programming, components and classification, industrial applications, and careers in robotics.

APPENDIX A. SOCIETIES AND ASSOCIATIONS

American Association for Artificial Intelligence
445 Burgess Drive, Menlo Park, CA 94025

Association for Unmanned Vehicle Systems
1133 15th Street NW, Washington, DC 20005

Automated Vision Association
P. O. Box 3724, 900 Victors Way, Ann Arbor, MI 48106

British Robot Association
28-30 High Street, Kempston, Bedford MK42 7BT, England

Homebrew Robotics Club
c/o Richard Prather, 91 Roosevelt Circle, Palo Alto, CA 94306
The organization has several active chapters.

IEEE Robotics and Automation Council
Institute of Electrical and Electronics Engineers
345 East 47th Street, New York, NY 10017

The Institute of Industrial Engineers
25 Technology Park, Norcross, GA 30092

International Biometric Association
1001 Connecticut Avenue NW, Suite 800, Washington, DC
20036

Japan Industrial Robot Association (JIRA)
Kikai Shinko Building 3-5-8, Shiba-Koen, Minato-ku, Tokyo, 105, Japan

Machine Vision Association
P. O. Box 930, One SME Drive, Dearborn, MI 48121

Marine Technology Society
1825 K Street NW, Suite 203, Washington, DC 20006

The Society has a committee on artificial intelligence and robotics.

National Personal Robot Association
P. O. Box 1366, Dearborn, MI 48121

Established in 1984 by the Robotics Industries Association for persons and corporations interested in or involved in the personal robot industry.

National Service Robot Association
P. O. Box 3724, Ann Arbor, MI 48106

An organization for persons and corporations interested in or involved in the application of robot technology to human services such as health care, education, security, space and undersea exploration, and related nonmanufacturing areas.

Robotic Institute of America
P. O. Box 930, Dearborn, MI 48128

The Robotic Society of America
P. O. Box 54-H, Scarsdale, NY 10583

Robotics Experimenters Amateur League (REAL)
P.O. Box 3227, Seal Beach, CA 90740

A nationwide organization with groups in several states.

Robotics Industries Association (RIA)
P.O. Box 3724, 900 Victors Way, Ann Arbor, MI 48106

Founded in 1974, the association is organized specifically to promote the use and development of robotic technology.

Publishes *Robot Times*, proceedings of meetings, books, directories, and a glossary.

Robotics Interest Group (ROBIG)
3205 Sydenham St., Fairfax, VA 22021
A personal robotics interest group.

Robotics International
Society of Manufacturing Engineers, P. O. Box 930, One SME Drive, Dearborn, MI 48121
Founded in 1980 for persons interested in or working with industrial robots.

Robotics Society of America
36 Newell Street, San Francisco, CA 94133
Chapters of this organization have been formed in many states. Sponsors a newsletter and PC-based robotics bulletin board.

United States Robotics Society
616 University Ave., Palo Alto, CA 94301
For the "garage" experimenter.

APPENDIX B. PERIODICALS, JOURNALS, AND NEWSLETTERS

Advanced Manufacturing Technology
was *Industrial Robots International*
Technical Insights, Inc., 32 North Dean Street, Englewood, NJ 07631.

Advances in Automation and Robotics, 1985-
JAI Press, Inc., 55 Old Post Road, No. 2, Box 1678, Greenwich, CT 06836-1678. Irregular.

Automation News: The Newspaper for the Robotics Industry
Grant Business Publications. c/o Friedberg, 450 Park Avenue, No. 2702, New York, NY 10222-2605.

Canadian Institute for Advanced Research in Artificial Intelligence and Robotics, 1988-
Ablex Publishing Corporation, 355 Chestnut Street, Norwood, NJ 07468. Annual.

Hobby Robot Design Notes
Hobby Robot Company, P. O. Box 997, Lilburn, GA 30247. Quarterly.

Experimental robotics.

IEC Report: A Newsletter on Japan Robotics
AAT Publishing, Inc., 16322 Manhattan Place, No. 1, Gardena, CA 90247. Monthly.

Reports on Japanese robotics developments.

IEEE Transaction on Robotics and Automation, 1985-
IEEE Robotics and Automation Council, 345 East 47th Street, New York, NY 10017-2394. Bi-monthly.
Technical aspects.

The Industrial Robot, 1973-
International Fluidics Services, Ltd., (IFS Publishing Company) 39 High Street, Kempston, Bedford MK47 7BT, England. Quarterly.
Provides international coverage of the robotics industry.

The International Journal of Robotics Research, 1982-
The MIT Press, Massachusetts Institute of Technology, 55 Hayward Street, Cambridge, MA 02142. Bi-monthly.
Technical aspects of robotics.

International Logo Exchange, 1986-
P. O. Box 5341, Charlottesville, VA 22905. Bi-monthly.
Covers uses of Logo in the classroom.

International Robotics Product Database, 1981-
TecSpec, Box 617024, Orlando, FL 32861-7024.

Journal of Robotic Systems, 1984-
John Wiley & Sons, Inc., Wiley Interscience Journals, 605 Third Avenue; New York, NY 10158. Quarterly.
Covers research reports on all aspects of robotics.

Logo and Education Computing Journal, 1983-
Interactive Education Foundation, 7 Flowerfield Building, St. James, NY 11780-1502. 5 per year.
Includes articles on robots, especially turtle robots.

Manufacturing Automation News
was *Robot News International*, 1981-
I.F.S. (Publications) International, 35-39 High Street, Kempston, Bedford MK42 7BT, England. Monthly.
Covers robotics industry news in England.

NSRA News
National Service Robot Association, P. O. Box 3724, Ann Arbor, MI 48106. Quarterly.

Newsletter provided for members of the National Service Robot Association. Covers happenings concerning service robots.

Personal Robotics Magazine
P. O. Box 421, Rheen Valley, CA 94570

Personal Robotics News, 1983-
Kensington Publishing Company, 6300 Telegraph Avenue, Oakland, CA 94609.

Provides information for manufacturers and vendors of personal robotics hardware.

Robomatix Reporter, 1983-
E I C-Intelligence, 48 West 38th Street, New York, NY 10018. Monthly.

Provides document summaries of publications pertaining to the robotics industry.

Robomatix Reporter Annual
Bowker A & I Publishing, 245 West 17th Street, New York, NY 10011.

Robot Insider
Fairchild Publications, 11 East Adams Street, Suite 1400, Chicago, IL 60603. Monthly.

Newsletter covers domestic and international robot industry.

Robot Reader, 1986-
P. O. Box 3243, El Paso, TX 79923-3243. Monthly.

Covers the personal robotics field and includes applications in education.

Robot Times
Robotic Industries Association, One SME Drive, Box 1366, Dearborn, MI 48121. Quarterly.

Provided for the members of the Robotic Institute of America. Covers information on conferences, workshops, symposiums, and publications.

Robot/X News, 1982-
Robotics Publications, Inc., Box 450, Mansfield, MA 02048. Every 2 months.

Covers news and articles for robotics industry management.

Robotica, 1983-
Cambridge University Press, Edinburgh Building, Shaftesbury Road, Cambridge CB2 2RU, England, and 32 East 57th Street, New York, NY 10022. Quarterly.

Includes articles and news on robotics, robotics research, and educational programs.

Robotics, 1985-
Elsevier Science Publishing Company, Inc. (New York), 52 Vanderbilt Avenue, New York, NY 10017. 4 per year.

Covers information about the international robotics industry, particularly applications in business and technological fields.

Robotics Age, 1983-1985.
Robotics Age, Inc., Strand Building, 174 Concord Street, Peterborough, NH 03458. Monthly.

Included articles on robotic technology developments and applications. Succeeded by *Robotics Engineering*.

Robotics Age Newsletter, 1983-
21 Century Media Communications, Inc., 3249 West 34th Avenue, Vancouver, BC V6N 2K3 Canada. Monthly.

Robotics and Automation, 1985-
International Association of Science and Technology for Development, Acta Press, 2990 Red Hill Road, Costa Mesa, CA 92626. 3 issues a year.

Technical aspects.

Robotics and Computer-Integrated Manufacturing, 1984-
Pergamon Press, Maxwell House, Fairview Park, Elmsford, NY 10523. Quarterly.

Includes papers on theoretical, experimental, and applied robotics.

Robotics and Expert Systems, 1985-
Instrument Society of America, 67 Alexander Drive, Box 122777, Research Triangle Park, NC 27709. Annual.

Robotics Engineering, 1985-1986. (Ceased publication with the December 1986 issue.)
Robotics Age, Inc., Strand Building, 174 Concord Street, Peterborough, NH 03458. Monthly.

Provided news and articles on industrial, commercial, and other applications of robots.

Robotics Patents Newsletter
Communications Publishing Group, Inc., 309 West Central Street, Suite 226, Natick, MA 01760. Monthly.

Reports on recently published United States and international patent documents in the field of robotics.

The Robotics Report; the Authoritative Washington Report for the Robotics Industry, 1981-
Washington Capital News Reports, Inc., 7043 Wimsatt Road, Springfield, VA 22151-4070. Monthly.

Reports on industrial robotics.

Robotics Research, 1982-
The MIT Press, 28 Carleton Street, Cambridge, MA 02142. Quarterly.

Robotics Technical Committee - Newsletter
Robotics Technical Committee, IEEE Computer Society, IEEE, Inc., 1730 Massachusetts Avenue NW, Washington, DC 20036-1903. Quarterly.

Reports on conferences, book reviews, job opportunities, recent developments, and other activities in the area of robotics.

Robotics Today, 1979-? (Ceased publication.)
Society of Manufacturing Engineers, Robotics International, One SME Drive, Box 930, Dearborn, MI 48121. Bi-monthly.

Covered applications of robotics in automated manufacturing (industrial robots), and included a "popular robotics" section which covered education, personal, and promotional robot systems.

Robotics World, 1983-
Communications Channels, Inc., 6255 Barfield Road, Atlanta, GA 30328. 6 times a year.

Covers articles and news of the robotics industry.

Robots, 1982-
F. A. Jour, 11 Rue du marche St. Honore, 75001 Paris, France. Monthly.

Covers robotics industry. Text in French and English.

TEC Report: A Newsletter on Japan Robotics
AAT Publishing, Inc., 16322 Manhattan Place, No. 1
Gardenia, CA 90247. Monthly.

Covers Japanese robot development and marketing.

See also periodicals dealing with artificial intelligence, which usually contain material on robots and robotics.

APPENDIX C. ROBOT MANUFACTURERS AND DISTRIBUTORS

21st Century Robotics
1790 Corporate Drive
Suite 330
Norcross, GA 30093

Acorn Computers Corporation
400 Unicorn Park Drive
Woburn, MA 01801

Amarobot
1780 Shattuck Avenue
Berkeley, CA 94709

Armatrol, Inc.
1780 Shattuck Avenue
Berkeley, CA 94709

American Micro Products
705 North Bowser
Richardson, TX 78081

Amptronics
Division of TII Corporation
401 North Salem Avenue
Arlington Heights, IL 60005

Analog Micro Systems
5660 Valmont Road
Boulder, CO 80301

Androbot, Inc.
550 Charcot Avenue
San Jose, CA 95131-9120

Android Amusement Corporation
1408 East Arrow Highway
Irwindale, CA 91706

Android Industries, Inc.
P. O. Box 1240
Morgan Hill, CA 95037

Arctec Systems, Inc.
9104 Red Branch Road
Columbia, MD 21045

Artur Fischer (UK), Ltd.
Fischer House
25 Newtown Road
Marlow, Bucks SL7 1JY
England

Axlon
P. O. Box 306
125 Main Street
Half Moon Bay, CA 94109

Bingel Robotics
P. O. Box 14655
Gainesville, FL 32604

Career Aids, Inc.
20417 Nordhoff Street
Chatsworth, CA 91311

Chessel Microproducts
Southdownview Road, Worthing
West Sussex BN14 8NL
England

Cincinnati Milacron
470 Marburg Avenue
Cincinnati, Ohio 45209
(For information on industrial robots)

Clwyd Technics
Antelope Industrial Estate
Rhydymwyn, Clwyd
Wales

Colne Robotics
207 Northeast 33rd Street
P. O. Box 23416
Fort Lauderdale, FL 33334
and
Beaufort Road/Richmond Road
East Twickenham, Middlesex
TW1 2PQ
England

Commotion Robotics, Ltd.
241 Green Street
Enfield, Middlesex EN3 7SJ
England

Comptech Systems Design
1722 Vermillion Street
Hastings, MN 55033

Compusa Corporation
Mountainside, NJ

ComRo Inc.
126 East 64th Street
New York, NY 10021
and
c/o Southwest Advisory, Inc.
2425 East Thomas Road
Suite 8
Phoenix, AZ 85016

Creative Learning Systems International
Trans Tech Systems Division
9889 Hibert Street
Suite E
San Diego, CA 92131

Cyber Robotics, Ltd.
Tilling Drive
Walton, Stone, Staffs
ST15 0SA
England

Cybernetic Applications
Portway Trading Estate
Andover, Hants
England

Cyberotics, Inc.
99 West Street
Medfield, MA 02052

Cybot, Inc.
12510 128th Avenue, Northeast
B-5
Kirkland, WA 98034

D & M Computing, Inc.
Box 2102
Fargo, ND 58102
and
Moorhead, MI

Droid Systems Corporation
412 East 88th Street
New York, NY 10128

E&L Instruments
70 Fulton Terrace
P.O. Box 1942
New Haven, CT 06509

Eastern Machinery & Manufacturing
Robotics Division
850 South West Temple
Salt Lake City, UT 84101

Economatics, Ltd.
4 Orgreave Crescent
Dore House Industrial Estate
Handsworth, Sheffield
S13 9LQ
England

Educational Resources
2360 Hassell Road
Hoffman Estates, IL 60195

Energy Concepts, Inc.
3254 North Kilbourn Avenue
Chicago, IL 60641

Eshed Robotec, Robot Technology, Inc.
Commercial Center
Ramat Ilan
P. O. Box 28346
Tel Aviv 61282
Israel

EZ Mower, Inc.
P. O. Box 77018
Houston, TX 77215

Feedback, Inc.
620 Springfield Avenue
Berkeley Heights, NJ 07922

Feedback Instruments
Park Road
Crowborough, East Sussex
England

fischer america, inc.
175 Route 46 West
Fairfield, NJ 07006

Flexible Systems
c/o Harvard Associates, Inc.
260 Beacon Street
Somerville, MA 02143

Flexible Systems
219 Liverpool Street
Hobart, Tasmania, Australia

Fordham Radio
260 Motor Parkway
Hauppauge, NY 11788

Frank Hogg Laboratory
770 James Street
Suite 215
Syracuse, NY 13203

General Development Company
312 East Breaburn Street
Phoenix, AZ 85022
(Logsdon 1984)

General Robotics Corporation
14618 West 6th Avenue
Suite 150
Golden, CO 80401

Genesis Computer Corporation
P. O. Box 152
Hellertown, PA 18055

Graymark
Box 5020
Santa Ana, CA 92704

Harvard Associates, Inc.
260 Beacon Street
Somerville, MA 02143

Hearlihy & Company
714 West Columbus Street
P. O. Box 869
Springfield, OH 45501

Heath Company
Benton Harbor, MI 49022

Heathkit/Zenith Educational Systems Division
Veritechnology Electronics Corp.
Department 570-445
Hilltop Road
St. Joseph, MI 49085

Hob-Bots of Hudson
P. O. Box 640
Hudson, OH 44236

Hobby Robot Company
P. O. Box 887
Hazelhurst, GA 31539
(also P.O. Box 997
Lilburn, GA 30247

Hubotics
5375 Avenida Encinas
Carlsbad, CA 92008

Ideal Division, CBS Toys, Inc.
183 Madison Avenue
New York, NY 10016

International Robotics
611 Broadway
Suite 422-B
New York, NY 10012

Iowa Precision Robotics, Inc.
908 10th Street
Milford, Iowa 51351

Isaac Robotics
P. O. Box 138
Old Chatham, NY 12136

Kelvin Electronics
7 Fairfield Avenue
Plainview, NY 11803

Kuhn & Schreiber, Inc.
2756 Lake Road
Hilton, NY 14468

Lab-Volt Systems
P. O. Box 686
Farmingdale, NJ 07727

LEGO Educational Department
P. O. Box 39
Enfield, CT 06030

Lennox Education Products
P. O. Box 809000
Dallas, TX

L. J. Electronics, Ltd.
1886 Larchwood
Troy, MI 48083
and
Francis Way
Bowthorp Industrial Estate,
Norwich NR5 9JA
England

Mack Corporation
3695 Industrial Drive
P. O. Box 1756
Flagstaff, AZ 86002

Maxtrom
1825A Durfee Avenue
South El Monte, CA 91733

Micro Management Systems, Inc.
2803 Thomasville Road East
Cairo, GA 31728

Microbot, Inc.
453-H Ravendale Drive
Mountain View, CA 94043

Microhmega
6918 16th Avenue, Northwest
Calgary, Alberta
Canada T2M0K3

Milton Bradley Company
443 Shaker Road
East Longmeadow, MA 01028

Mitsubishi Electric Sales America, Inc.
799 North Bierman Circle
Mount Prospect, IL 60056

Mitsubishi (UK), Ltd.
Hertford Place, Denham Way
Maple Cross, Rickmansworth
WS3 2BJ
England

MJR Robotics
Brook House
Priors Leaze Lane
Hambrook Nr Chichester,
West Sussex
England

Mondo-Tronics
20090 Rodrigues Avenue, #1
Cupertino, CA 95014

Multibotics
Access Software, Inc.
2561 South 1560 West
Woods Cross, UT 84087

Nationwide Robots
4141 Woerner Avenue
Levittown, PA 19057

New Tech Promotions
2265 Westwood Boulevard
Suite 248
Los Angeles, CA 90024

Nintendo
4820 150th Avenue Northeast
P. O. Box 957
Redmond, WA 98052

Northern Computers
Education and Training
Division
Churchfield Road
Frodsham, Cheshire
WA6 6RD
England

OWI Inc.
1160 Mahalo Place
Compton, CA 90220

Parsec Research
41805 Albrae Street
Fremont, CA 94538

Personal Robotics
Corporation
469 Waskow Drive
San Jose, CA 95123

Play Jour, Inc.
200 Fifth Avenue
Suite 1024
New York, NY 10010

Powertran Cybernetics
Portway Industrial Estate
Andover, Hants SP10 3NN
England

Prep, Inc.
1007 Whitehead Road
Extension
Trenton, NJ 08683

Prism Micro Products, Ltd.
Prism House
18/29 Mora Street, City Road
London EC1V 8BT
England

Quasar Industries, Inc.
The Robot People
59 Meadow Road
Rutherford, NJ 07070
(Display robots - Klatu)

Rebs Robots
3530 Timmons Lane, #155
Houston, TX 77027

Reekie Research Company, Ltd.
Beaufort Works
Beaufort Road/Richmond Road
East Twickenham, Middlesex
TW1 2PH
England

Remcon Electronics
P. O. Box 148258
Nashville, TN 37214
and
P. O. Box 81
Chislehurst, Kent BR7 6LP
England

Rhino Robots, Inc.
3402 North Mattis
P. O. Box 4010
Champaign, IL 61820

Rio Grande Robotics
1220-A North Third Street
Las Cruces, NM 88005

Robot City Technology
20 Burners Lane
Kiln Farm, Milton Keynes
MK11 3HB
England

Robot Company
2189 Northwest 53rd Street
Fort Lauderdale, FL 33309

The Robot Factory
P. O. Box 112
Cascade, CO 80809

Robot Mart
19 West 34th Street
New York, NY 10001

Robot Repair
816 1/2 21st Street
Sacramento, CA 95814

Robot Shack
See Robot Shop

Robot Shop
P. O. Box 582
El Toro, CA 92630

RobotEx
111 East Alton Avenue
Santa Ana, CA 92707

Robotics International Corporation
2335 East High
Jackson, MI 49203

Robotland
4251 North Federal Highway
Boca Raton, FL 33431

Robotorium
252 Mott Street
New York, NY 10012

Robotronics
229 East 52nd Street
New York, NY 10022

Sandhu Machine Design, Inc.
3402 North Mattis Avenue
Champaign, IL 61320

Servitron Robots, Inc.
1009 Grant Street
Denver, CO 80203

ShowAmerica, Inc.
841 North Addison Avenue
Elmhurst, IL 60126

Spectron Instrument Corp.
1342 West Cedar Avenue
Denver, CO 80223

LW Staines & Company
Unit 2
Roding Industrial Estate
London Road
Barking, Essex IG11 8BU
England

Stock Model Parts
54 South Denton Avenue
New Hyde Park, NY 11040

Swisscomp, Inc.
Tampa, FL

Syke Instrumentation Company, Ltd.
Syke House
117-119 Station Road
Liss, Hants GU33 7AJ
England

Systems Control
3A Highforce Road
Riverside Park Industrial Estate, Middlesborough TS2 1LW
England
and
c/o TecQuipment
Box 1074
Acton, MA 01720

Tandy Corporation
1500 One Tandy Center
Fort Worth, TX 76102

Teachers' Laboratory, Inc.
Robocar
P. O. Box 6480
Brattleboro, VT 05301

Technical Micro Systems, Inc.
366 Cloverdale Street
P. O. Box 7227
Ann Arbor, MI 48105

Technovate, Inc.
910 Southwest 12th Avenue
Pompano Beach, FL 33060

TecQuipment International, Ltd.
Bonsall Street
Long Eaton, Nottingham NG10 2AN
England

Terrapin, Inc.
400 Riverside Street
Portland, ME 04103

TII Robotic Systems
Palatine, IL

Tomy Corporation
901 East 233rd Street
P. O. Box 6252
Carson, CA 90748

Transtech Systems
(See Creative Learning Systems)

TTC Corporation
2009 East 233rd Street
Carson, CA 90810

Unimation, Inc.
Shelter Rock Lane
Danbury, CT 06810
(For information on industrial robots)

Universal Machine Intelligence
The Royal Victoria Patriotic Building
Trinity Road
London SW18 3SX
England

Universal Robot Systems
7025 El Paseo Street
Long Beach, CA 90815

Valiant Designs, Ltd.
Park House
140 Battersea Park Road
London SW11 4NB
England

Valley Robotics
2201 The Alameda
Suite 10
Santa Clara, CA 95050
(Educational robot supplier)

World of Robots
2335 East High Street
Jackson, MI 49203

APPENDIX D. EDUCATIONALLY USEFUL ROBOTS

ROBOT ARMS

Alfred (Robot City Technology)
Alpha I (Microbot, Inc.)
Alpha II+ Robot Arm System (Microbot, Inc.)
Armatrol (Feedback, Inc.)
Armatron (Tandy Corporation; Radio Shack)
Armborg (Feedback, Inc.)
Armdraulic I (Feedback, Inc.)
Armdraulic II (Feedback, Inc.)
Armdroid I (Colne Robotics; D & M Computing, Inc., distributor)
Armdroid II (Colne Robotics, Inc.; D & M Computing, Inc., distributor)
Armover (Feedback, Inc.)
Armsort (Feedback, Inc.)
Armtek I (Energy Concepts, Inc.)
Atlas (L J Electronics, Ltd.)
Atlas (Technovate, Inc.)
BASE Series II (Mack Corporation)
BASE Series II WR (Mack Corporation)
Beasty Robot Arm (Commotion Robotics)
Biometal DH-101 (Toki Corporation; Valley Robotics)
BRAT 501 (Amptronics)
BRAT 801 (Amptronics)
Capek I (FMS International, Inc.)

Cepek 1 - SCARA - Selective Compliance Assembly Robot Arm (Reekie Research Co., Ltd.)
Centari (Amatrol, Inc.)
Cyber 310 FORTH (Cyber Robotics, Inc.; Acorn Computers Corporation)
Dissectible Robot System (Lab-Volt Systems)
Explorer (TII Robotic Systems)
fischertechnik Robot Training Arm (Valley Robotics, distributor,kit)
Genesis P 101 (Powertran Cybernetics
Genesis P 102 (Powertran Cybernetics)
Herbot II (Powertran Cybernetics)
Hercules (Amatrol, Inc.)
HRA 933 - ready-built version of Genesis P 101 (Feedback Instruments)
HRA 934 - ready-built version of Genesis P 102 (Feedback Instruments)
IVAX - SCARA - Selective Compliance Assembly Robot Arm (Feedback, Inc.; Powertran Cybernetics)
Jupiter (Amatrol)
MA 2000 (TecQuipment International, Ltd.)
Mentor (Cybernetic Applications)
Mercury (Amatrol)
Micro Grasp (Powertran Cybernetics)
MiniBRAT (TII Robotic Systems)
MiniMover (Microbot, Inc.)
Minimover (Syke Instrumentation Company, Ltd.)
Mobile Armatron Robot Arm (Tandy Corporation; Radio Shack)
Model 5440 (Technovate, Inc.)
Movemaster I-RM-101 (Mitsubishi Electric Sales America; E & L Instruments; Robotex)
Movemaster II-RM-501 (Mitsubishi Electric Sales America, Inc.; Robotex)
Neptune (Cybernetic Applications)
Neptune II or Neptune 2 Hydro-Electric Robot (Cybernetic Applications; Alan-Hayes Corporation)
Ogre Robot Arm (LW Stains & Company)
Pedro - SCARA - Selective Compliance Assembly Robot Arm (Feedback Instruments)
Pneumatic Robot Training System (Lab-Volt Systems)

PZ 80 Micom (Hikawa; Compusa Corporation, distributor)
Reekie Arm (Reekie Research Company, Ltd.)
Rhino SCARA (Rhino Robots, Inc.)
Rhino XR-2 (Rhino Robots, Inc.)
Rob 3 (Swisscomp, Inc.)
Robot I (Automated Armatron; Analog Micro Systems)
Robot I (Spectron Instrument Corporation)
Robot IV (Spectron Instrument Corporation)
Robot Arm Trainer (Heath Company)
Robotic Arm II (Kuhn & Schreiber, Inc.; Hearlihy & Company, distributor)
Robotic Arm Kit (Hearlihy & Company, distributor)
Robotic Arm Kit (Kelvin Electronics)
Robotnick P/L (Harvard Associates, distributor)
RTX - modified SCARA - Selective Compliance Assembly Robot Arm (Universal Machine Intelligence)
Scorbot-ER III (Eshed Robotics, Ltd. Prep, Inc., distributor)
Serpent - SCARA - Selective Compliance Assembly Robot Arm (Cybernetic Applications)
SIR-1 Arm (Haikato Robotics. Rio Grande Robotics, distributor)
Smart Arm MA-83 (Systems Control)
TARA Robot (L. J. Electronics, Ltd.)
TeachMover Robotic Arm (Microbot, Inc.)
Teachmover (Syke Instrumentation Company, Ltd.)
Teach Robot (Remcon Electronics)
Toki Nitinol Arm (Valley Robotics)
Tutor (Cybot, Inc.)

ROBOTIC TURTLES

Armadillo (Feedback, Inc.)
Ernie (Microhmega)
F.R.E.D. (Androbot)
GR-15 (Chessel Microproducts, distributor)
ITSABOX; Itsa-Box (Technical MicroSystems, Inc.; kit)
Memocon Crawler (Movit Educational Robot Kits; See KITS)

Nomad (Frank Hogg Laboratory - TRS-80 color computer)
S-Cargo (Movit Educational Robot Kits; See KITS)
Tasman Turtle (Flexible Systems, c/o Harvard Associates, Inc.; Transtech Systems, distributors)
Turtle (Amarobot)
Turtle Tot (Flexible Systems, c/o Harvard Associates, Inc.; Rio Grande Robotics; Transtech Systems, distributors)
Valiant Turtle (Valiant Designs, Ltd. Harvard Associates, Inc.; Rio Grande Robotics, distributors)
WAO (Career Aids, Inc., distributor; See KITS)
Zeaker (D & M Computing, Inc., distributor)

WHEELED ROBOTS

Andy (Axlon)
ARES (Auxiliary Robotic Educational System) (21st Century Robotics)
ART-2 (Hobby Robot Company, Inc.)
BBC Buggy (Economatics, Ltd.)
Beasty (Commotion Robotics, Ltd.)
Beasty Mobile Base (Commotion Robotics, Ltd.)
Bob; BOB; B.O.B.; Brains on Board (Androbot, Inc.)
Canbot (The Robot Factory)
Charger (Rhino Robots, Inc.)
Circular Robot (Movit Educational Robot Kits; See KITS)
COM 701, COM 702, COM 703, COM 704, COM 705 (Comptech Systems Design
CompuRobot (Maxtron)
CompuRobot II (Maxtron)
ComRoToT (Comro, Inc.)
Cyber I (Cyberotics, Inc.)
Dingbot (Tomy Corporation)
ELAMI (Robotland)
Gemini (Arctec Systems)
GENESIS I (Droid Systems Corporation)
George (Armarobot)
Hero 1 (Heath Company; also available as kit)
Hero 2000 (Heath Company; also available as kit)

Hero Jr. (Heath Company; also available as kit. Rio Grande Robotics, distributor)
Hobbybot Robot (Alan Branch Associates. Rio Grande Robotics, distributor. Also available as kit from distributor)
Hot Tot on a Tricycle (The Robot Factory)
Hubot (Hubotics, Inc.)
K1 Buggy (MJR Robotics)
Line Tracer II (Movit Educational Robot Kit; See KITS)
Martha (Armarobot)
Marvin Mark I (Iowa Precision Robotics, Inc.)
Maxx Steele (Ideal Div., CBS Toys, Inc.)
MicroDroid (Hob-Bots of Hudson)
Micro-Kitten (Spectron Instruments)
Micro Pulse Buggy (Northern Computers)
The Milk Carton Robot (The Robot Factory)
Navious, Navis, Navius (Movit Educational Robot Kit; See KITS)
Nomad (Genesis Computer Corporation - Commodore computer)
Omnibot (Tomy Corporation)
Omnibot 2000 (Tomy Corporation)
Piper Mouse (Movit Educational Robot Kit; See KITS)
RB5X (General Robotics Corporation. Rio Grande Robotics, distributor)
Robie Jr. (Tandy Corporation; Radio Shack)
Robie Sr. (Tandy Corporation; Radio Shack)
Robocar I and II (Teachers' Laboratory, Inc.)
Robocycle (Robot Repair)
Robodroid (The Robot Factory)
Robomower (EZ Mower - robot lawn mower)
Robotic Training System (Lab-Volt Systems)
Rollin' Robie Robot (Tandy Corporation; Radio Shack)
RoPet (Personal Robotics Corporation)
Scorpion (Sandhu Machine Design, Inc.)
Scorpion (Rhino Robots, Inc.)
Sentroid (Hob-Bots of Hudson)
Six T. Robot (The Robot Factory)
Smart Rabbit Robot (Hobby Robot Company, Inc., kit)
TeachMover (Microbot, Inc.)

Topo (Androbot, Inc.)
Topo III (Androbot, Inc.)
TREKKER (Clwyd Technics)
Verbot (Tomy Corporation)
Videobot (The Robot Factory)
XR-2 (Microbot, Inc.)

LEGGED ROBOTS

Avoider (Movit Educational Robot Kit; See KITS)
Medusa (Movit Educational Robot Kit; See KITS)
Mr. Bootsman (Movit Educational Robot Kit; See KITS)
Sound Skipper (Movit Educational Robot Kit; See KITS)
Turnbacker (Movit Educational Robot Kit; See KITS)

KITS

Capsela 1000 (Play Jour, Inc.; Career Aids, distributor)
Capsela 2000 (Play Jour, Inc.; Career Aids, distributor)
Capsela 5000 (Play Jour, Inc.)
Dissectible Robot System (Lab-Volt Systems)
DROID-BUG (Robot Shop)
EPR1 Robot (Lennox Education Products)
fischertechnik Robot Training Arm (fischer america. Career Aids; Valley Robotics, distributors)
fischertechnik Robotic Computing Kit (Artur Fischer (UK) Ltd.; fischer america. Creative Learning Systems International; Educational Resources; Parsec Research; Rio Grande Robotics, distributors)
Hero 1 (Heath Company)
Hero 2000 (Heath Company)
Hero Jr. (Heath Company)
Hobbybot Robot (Alan Branch Associates. Rio Grande Robotics, distributor)
Hydrobot (Amarobot - water-powered)
Isaac (Amarobot)
Itsa-Box (Technical Micro Systems, Inc.)

K1 Buggy (MJR Robotics)
LEGO TC Logo (LEGO Educational Department)
Movit Educational Robot Kits
  Avoider (Career Aids, Inc.; Fordham Radio; Hearlihy & Company; OWI, Inc.; Prism Micro Products, Ltd.; Rio Grande Robotics, distributors)
  Circular Robot (Career Aids, Inc.; Hearlihy & Company; OWI, Inc.; Rio Grande Robotics, distributors)
  Line Tracer II (Fordham Radio; Hearlihy & Company; OWI, Inc.; Rio Grande Robotics, distributors)
  Medusa (Career Aids, Inc.; Fordham Radio; Hearlihy & Co.; OWI, Inc.; Rio Grande Robotics, distributors)
  Memocon Crawler (Career Aids, Inc.; Fordham Radio; Hearlihy & Company; OWI, Inc.; Rio Grande Robotics; Stock Model Parts, distributors)
  Mr. Bootsman (Fordham Radio; OWI, Inc.; Rio Grande Robotics, distributors)
  Navious (Rio Grande Robotics); Navis (Fordham Radio); Navius (OWI, Inc.), distributors
  Q-Bot Kit (Eastern Machinery and Manufacturing)
  Peppy (Career Aids, Inc.; Hearlihy & Company; OWI, Inc.; Rio Grande Robotics, distributors)
  Piper Mouse (Career Aids, Inc.; Fordham Radio; Hearlihy & Company; OWI, Inc.; Rio Grande Robotics, distributors)
  S-Cargo (Fordham Radio, distributor)
  Sound Skipper (Rio Grande Robotics, distributor)
  Turnbacker (Fordham Radio; OWI, Inc.; Rio Grande Robotics, distributors)
  WAO; Wise Argent Orb (Career Aids, Inc.; Hearlihy & Company; Rio Grande Robotics, distributors)
PZ 80 Micom Robotic Arm Kit (Hikawa. Compusa Corporation, distributor)
Robot Arm Trainer (Heath Company)
Robot Kits (Graymark, distributor)
  Line Tracer Robot (Movit Line Tracer II)
  Programmable Robot (Movit Memocon Crawler)
  Robot Navigator (Movit Navius)
  Robot Walker (Movit Avoider)
  Robot Wheel (Movit Circular Robot)
  Sound Operated Robot (Movit Peppy)

Robotic Arm Kit (Hearlihy & Company, distributor)
Robotic Laboratory (Similar to fischertechnik kit) (Parsec Research)
Robotix R-1000 (Milton Bradley Company)
Robotix R-2000 (Milton Bradley Company)
Scorpion Mobile Robot (Rhino Robots, Inc.)
Smart Rabbit Robot (Hobby Robot Company, Inc.)
Space Wings Robotics Kit (Mondo-Tronics)
Teach Robot (Remcon Electronics, distributor)
X-1 (Robot Shop)
Z-1 (Robot Shop)
Z-2 (Robot Shop)

COMPUTER-CONTROLLED PROJECT KITS

The Computing Construction Kit (fischer america)
Computing Experimental (fischer america)
Exploring the Industrial Applications for Robotics. Course to Accompany Robot Arm Trainer (Heath Company)
fischertechnik Training Robot (fischer america)
Multibotics MB230 Robotic Workshop and B100 Interface Unit - using Capsela modules (Multibotics)
Plotter/Scanner (fischer america)
Robocar I and II (Teachers' Laboratory, Inc.)
TSI Robotics Kits (Terrapin, Inc.)
 TSI - 1 Kit
 TSI - 2 Kit (with light sensors)
  (Use with Capsela, Erector, fischertechnik, Robotix, and Legos - Lego/Logo)

VISION SYSTEMS

Commotion Robot Vision - SNAP EV1 Camera (Commotion Robotics)

APPENDIX E. ROBOT PARTS AND ELECTRONICS SUPPLIERS

Active
P. O. Box 8000
Westborough, MA 01581

Advanced Computer Products
P. O. Box 17329
Irvine, CA 92713

All Electronics Corporation
905 South Vermont Avenue
P. O. Box 20406
Los Angeles, CA 90006

Alltronics
15460 Union Avenue
San Jose, CA 95124

Alpha Products
7904-N Jamaica Avenue
Woodhaven, NY 11421

American Design Components
62 Joseph Street
Moonachie, NJ 07074

Amarobot
2913 Ohio Street
Richmond, CA 94804

Analytic Methods
1800 Bloomsbury Avenue
Ocean, NJ 07712

B & B Motor and Control Corporation
Corporate Marketing and Communications
Route 4 and Covey Road
Burlington, CT 06013
(issues *Motion Control Handbook* on motors)

Ball Brothers Research Corporation
Electronic Display Division
1633 Terrace Drive
St. Paul, MN 55113

Barrett Electronics
5312 Buckner Drive
Lewisville, TX 75028

BCD Electro
P. O. Box 830119
Richardson, TX 75083-0119

Bishop Graphics, Inc.
5388 Sterling Center Drive
P. O. Box 5007
Westlake Village, CA 91359
(E-Z Circuit, adhesive etch patterns)

Winfred M. Berg
499 Ocean Avenue
East Rockaway, Long Island, NY 11518

Boston Gear
14 Hayward Street
Quincy, MA 02171

Byte-Back Company
Route 3, Box 147
Leesville, SC 29070

Chaney Electronics
P. O. Box 27038
Denver, CO 80227

Colne Robotics
207 North 33rd Street
Fort Lauderdale, FL 33334

Comprehensive Guides
7507 Oakdale Avenue
Canoga Park, CA 91306

Computer Parts Mart
3200 Park Boulevard
Palo Alto, CA 94306

Cybernetic Micro Systems
445-203 San Antonio Road
Los Altos, CA 94022

Deltroniks, Inc.
5151 Buford Highway
Atlanta, GA 30340
(Surplus equipment)

Digi-Key Corporation
Highway 32 South
P. O. Box 677
Thief River Falls, MN 56701

Dick Smith Electronics
P. O. Box 2249
Redwood City, CA 94064-2249

DoKay
2100 De la Cruz Boulevard
Santa Clara, CA 95050

Edmund Scientific Company
101 East Gloucester Pike
Barrington, NJ 08007
(Big Trak motors, some surplus supplies)

Electro-Craft Corporation
1600 Second Street South
Hopkins, MN 55343

Erac Company
8280 Clairemont Mesa Boulevard
Suite 117
San Diego, CA 92111

Gates Energy Products
1050 South Broadway
Denver, CO 80217

G. B. Micro
P. O. Box 280298
Dallas, TX 75228

GIL Electronics
P. O. Box 1628
Soquel, CA 95073

Globe Union Battery
5757 North Green Bay Avenue
Milwaukee, WI 53201

GTE Microcircuits
2000 West 14th Street
Tempe, AZ 85281
(Touch-tone ICs)

H & R Corporation
401 East Erie Avenue
Philadelphia, PA 19134
(Big Trak motors, surplus electrical and mechanical equipment)

Halted Specialties Company
827 East Evelyn Avenue
Sunnyvale, CA 94086

Hal-Tronix, Incorporated
12671 Dix-Toledo Highway
P. O. Box 1101
Southgate, MI 48195

Hewlett Packard Components
640 Page Mill Road
Palo Alto, CA 94034
(Optoelectronic components)

Hi-Tek Sales/BNF
119 Foster Street
Peabody, MA 01961-3357

Hobby Robot Company
P. O. Box 997
Lilburn, GA 30247

Inland Motors
Industrial Drives Division
Kollmorget Corporation
201 Rock Road
Radford, VA 24141

Intel Corporation
3065 Bowers Avenue
Santa Clara, CA 95051
(8748 single-chip micro)

Interstate Electronics Corporation
1001 East Ball
P. O. Box 3117
Anaheim, CA 91803

Jameco Electronics
1355 Shoreway Road
Belmont, CA 94002

JDR Microdevices
1224 South Bascom Avenue
San Jose, CA 95128

JERRYCO, Inc.
601 Linden Place
Evanston, IL 60202
(Surplus electronic and mechanical equipment)

Mark V Electronics
248 East Main Street
Suite 100
Alhambra, CA 91801

Marlin P. Jones & Associates
P. O. Box 12685
Lake Park, FL 33403
(Surplus electronics)

Micro Switch
11 West Spring Street
Freeport, IL 61032

Microprocessors, Unlimited, Inc.
24,000 South Peoria Avenue
Beggs, OK 74421

Mitel Semiconductor
P. O. Box 13320
360 Legget Drive
Kanta, Ontario, Canada K2K1X5
(Touch-tone ICs)

MJR Digital
Mason Road
Milford, NH 03055
(Encoder disks)

Motorola Semiconductor Products, Inc.
3501 Ed Bluestein Boulevard
Austin, TX 78721

Motorola Semiconductors
Box 20912
Phoenix, AZ 85036

Mouser Electronics
11433 Woodside Avenue
Santee, CA 92071;
2401 Highway 287
North Mansfield, TX 76063

National Semiconductor Corporation
2900 Semiconductor Drive
Santa Clara, CA 95051

Plessey Semiconductor
Plessey Solid State
3 Whatney
Irvine, CA 92714

Polaroid Corporation
Commercial Battery Division
784 Memorial Drive
Cambridge, MA 02139
(Ultrasonic Ranging System Designer's Kit for sonar sensors)

Priority One Electronics
9161 Deering Avenue
Chatsworth, CA 91311

Radio Shack stores
or Radio Shack
One Tandy Center
Fort Worth, TX 76102

R & D Electronics
1202H Pine Island Road
Cape Coral, FL 33909

R + D Electronic Supply
100 East Orangethorpe Avenue
Anaheim, CA 92801

R. F. Electronics
1086-C North State College Boulevard
Anaheim, CA 92806

The Robot Mart
Room 1113
19 West 34th Street
New York, NY 10001

Robot Repair
816 1/2 21st Street
Sacramento, CA 95814

Robot Shop
P. O. Box 582
El Toro, CA 92630

Robotronics
229 East 52nd Street
New York, NY 10022

RS Components, Ltd.
P. O. Box 99
Corby, Northants NN17 9RS
England

SBC Mart
P. O. Box 1296
Ridgecrest, CA 93555

Sharon Industries
693 Brokaw Road
San Jose, CA 95112

Signetics
811 East Arques Avenue
Sunnyvale, CA 94016
(Motor drivers, servo systems, speed controls)

Silconix
2201 Laurelwood Road
Santa Clara, CA 95054

Silicon Valley Surplus
4401 Oakport
Oakland, CA 94601

Small Parts Inc.
6901 N. E. Third Avenue
P. O. Box 381736
Miami, FL 33138
(Gears and other precision parts)

Sprague Electric
70 Pembroke Road
Concord, NH 03301
(Stepper drivers, Hall-effect ICs, motor drivers)

Standard Microsystems Corporation
35 Marcus Boulevard
Hauppauge, NY 11787
(Communications circuits)

Stock Drive Products
55 South Denton Avenue
New Hyde Park, NY 11040
(Issues the *Handbook of Small Standardized Components*.)

Superior Electric Company
383 Middle Street
Bristol, CT 06010

Surplus Center
1015 West "O" Street
P. O. Box 82209
Lincoln, NE 68501
(Surplus equipment)

Technomatic, Ltd.
17 Burnley Road
London NW10 1ED
England

Texas Instruments
P. O. Box 225012
Dallas, TX 75265
(Motor drivers, Hall-effect sensors)

Valley Computers
613 North Idlewild
Kaukauna, WI 54130

Vestra Technology, Inc.
7100 West 44th Avenue
Suite 101
Wheatridge, CO 80033

W. S. Jenks & Son
2024 West Virginia Avenue NE
Washington, DC 20002

Windsor Distributors
19 Freeman Street
Newark, NJ 07105

Winfred M. Berg Inc.
499 Ocean Avenue
East Rockaway, Long Island, NY 11518
(Gears and other precision parts)

Zilog
1315 Dell Drive
Campbell, CA 95008

APPENDIX F. NONPRINT MEDIA PRODUCERS AND DISTRIBUTORS

Agency for Instructional Technology
Box A
Bloomington, IN 47402

AIMS Media
6901 Woodley Avenue
Van Nuys, CA 91406

BBC Enterprises Film and Video Library
The Langham, Portland Place
London W1A 1AA, England

Bergwell Productions
106 Charles Lindbergh Boulevard
Uniondale, NY 11553

Brainworks, Inc.
24009 Ventura Boulevard
Calabasas, CA 91302

California State University
Center for Instructional Media
Sacramento, CA

Career Aids, Inc.
20417 Nordhoff Street
Chatsworth, CA 91311

The Center for Humanities, Inc.
Communications Park
Box 1000
Mount Kisco, NY 10549-9989

Charles Clark Company, Inc.
905 South 2nd Street
Ronkonkoma, NY 11779

Children's Video Library
60 Long Ridge Road
P. O. Box 4995
Stamford, CT 06907

Coronet Film & Video
420 Academy Drive
Northbrook, IL 60062

Coronet Films & Video
108 Wilmot Road
Deerfield, IL 60015

Deltak, Inc.
1751 Diehl Road
Naperville, IL 60566

Document Associates, Inc.
1697 Broadway
Suite 802
New York, NY 10019

Document Associates, Inc.
211 East 43rd Street
New York, NY 10017 (1986)

Educational AudioVisual, Inc.
17 Marble Avenue
Pleasantville, NY 10570

Educational Dimensions Group
Stamford, CT

Encyclopaedia Britannica Educational Corporation
425 North Michigan Avenue
Chicago, IL 60611

Epyx
1043 Kiel Court
Sunnyvale, CA 94089

Excalibur Technologies Corporation
800 Rio Grande Boulevard, Northwest
Albuquerque, NM 87104
(personal robot computer software)

Eye Gate Media
Chicago, IL

Films Incorporated
5547 North Ravenswood
Chicago, IL 60640-1199

Fullerton College
321 Chapman Avenue
Fullerton, CA 92634

GPN
P. O. Box 80669
Lincoln, NE 68501

Guidance Associates
Communications Park
Box 3000
Mount Kisco, NY 10549-0900

Hearlihy & Company
714 West Columbia Street
Springfield, OH 45501

Hearst Metrotone News
235 East 45th Street
New York, NY 10017

Human Relations Media
175 Tompkins Avenue
Pleasantville, NY 10570

Johnson/Rudolph Educational Resources
A/V Distribution Center
1027 Broadway
Bowling Green, KY
42104-9941

Journal Films, Inc.
Journal Video, Inc.
930 Pitner Avenue
Evanston, IL 60202

Karol Media
22 Riverview Drive
Wayne, NJ 07470-3191

King Features Entertainment
235 East 45th Street
New York, NY 10017

The Learning Company
545 Middlefield Road
Menlo Park, CA 94025

Library Filmstrip Center
Division of Meridian Education Corp.
205 E. Locust Street
Bloomington, IL 61701

McGraw-Hill Films
674 Via de la Valle
P. O. Box 641
Del Mar, CA 92014

McGraw-Hill Films, Webster Division
1221 Avenue of the Americas
New York, NY 10020

Media Basics Video
Larchmont Plaza
Larchmont, NY 10538

The MIT Press
28 Carleton Street
Cambridge, MA 02142

National Audiovisual Center
General Services Administration
Washington, DC 20409

National Geographic Society
Educational Services
Dept. 85
Washington, DC 20036

Prentice-Hall Media
See Vocational Media Associates

Prismatron Productions, Inc.
155 Buena Vista Avenue
Mill Valley, CA 94941

Pyramid Film & Video
Box 1048
Santa Monica, CA 90406

Random House School Division
Dept. 9050
400 Hahn Road
Westminster, MD 21157

Robotronix, Inc.
P. O. Box 1125
Los Alamos, MN 87544
(personal robot computer software)

Scholastic Software
Scholastic Inc.
2931 East McCarty Street
P. O. Box 7502
Jefferson City, MO 65102

Science Screen Report
P. O. Box 691
New York, NY 10003

Sigler Software
P. O. Box 3010
Charlottetown, PEI
Canada C1A 7N9

Society for Visual Education, Inc.
1345 Diversey Parkway
Chicago, IL 60614-1299

Society of Manufacturing Engineers
One SME Drive
Dearborn, MI 48121

Solana Studios
P. O. Box 663
4143 North 27th St.
Milwaukee, WS 53201

Sunburst Communications
39 Washington Avenue
Pleasantville, NY 10570-9971

Unicorn Educational Software
2950 East Flamingo Road
Suite B
Las Vegas, NV 89121

University of California Extension Media Center
2223 Fulton Street
Berkeley, CA 94720

The Video Schoolhouse
2611 Garden Road
Monterey, CA 93940

Vocational Media Associates
Box 1050
Mount Kisco, NY 10549-0050

WNET-TV
New York, NY

INDEX